LANDSLIDES: CAUSES, CONSEQUENCES & ENVIRONMENT

LANDSLIDES:

causes, consequences
& environment

Michael J. Crozier

CROOM HELM
London • Sydney • Dover, New Hampshire

British Library Cataloguing in Publication Data

Crozier, Michael J.
    Landslides: causes, consequences and
    environment.
    1. Landslides – New Zealand
    I. Title
    551.3'03'09931   QE599.N45
    ISBN 0-7099-0790-7

Croom Helm, 51 Washington Street, Dover,
New Hampshire 03820, USA

Library of Congress Cataloging in Publication Data

Crozier, Michael J.
    Landslides; causes, consequences, and environment.

    Bibliography: p.
    Includes index.
    1. Landslides. I. Title.
QE599.A2C76   1986   551.3   85-29048
ISBN 0-7099-0790-7

Printed and bound in Great Britain by
Biddles Ltd, Guildford and King's Lynn

CONTENTS

Contents

TABLES

FIGURES

To my father
who lifted my eyes to the hills

PREFACE

A satisfactory explanation of the landslide phenomenon would be one that provided answers to a number of basic questions: what are landslides; how do they function, where do they occur and what determines their location; when do they occur and what controls their timing; what is their human and environmental significance? Unfortunately, the traditional separation and rigid boundaries of disciplines have meant that many of these questions have been treated in isolation.

It is only in the last few years that the contributions from many diverse disciplines have begun to come together to enhance our understanding of the landslide problem. Multidisciplinary themes have been introduced at some conferences and more recently a few multi-authored publications have emerged. Landslides: Causes, Consequences and Environment is an attempt by a single author to continue this process of synthesis. The advantage for the single author over the editor of a multi-authored publication is the facility to formulate, not only the answers given, but, more importantly, the questions asked.

The kind of questions I have posed in this book inevitably reflect my background and perspective. As a student studying natural sciences in New Zealand, it was easy to be impressed with the importance of mass movement in the local landscape: landslides tore chunks from the mountains, created lakes, turned rivers to 'chocolate', and stripped soil from the farms. It is little wonder that Denys Brunsden, a visiting English geomorphologist, on his first view of New Zealand mountains, stated that 'New Zealand has the finest soil erosion in the world'. Despite the embarrassment to soil conservators when these words appeared as headlines in the Christchurch Press, I knew exactly what he meant.

However, geomorphology text books of the day, coming mostly from the Northern Hemisphere, did not seem to share the same enthusiasm for the role of mass movement; instead they were preoccupied with fluvially fashioned landscapes. While outside the lecture room walls, New Zealand went 'down to the sea in slips' we, inside, debated the fine-points of down-wasting of slopes versus backwearing - or was it really slope replacement!

However, not all New Zealand academics of this time had a blind-spot for landslides. While Sir Charles Cotton was feasting his magnificent interpretive vision on the landscape (albeit through Davisian spectacles), Professor Benson was writing one of the finest geological interpretations of regional slope instability ever published (Benson 1940).

At the same time, the practioners (civil engineers, agricultural engineers and soil conservators) were making steady progress with the landslide problem and building their own body of knowledge. In the English speaking world, geomorphologists and geologists gained access to these advances mainly through the lucid writing and generous interest of people such as Professor Skempton, Professor Hutchinson and Karl Terzaghi. The information proffered was quickly and fruitfully applied to geomorphic questions by authors such as Michael Carson, Michael Kirkby, Denys Brunsden and Michael Selby.

A number of others were responsible for fostering the study of landslides as a systematic science in its own right. Notable amongst these are the Americans Charles Sharpe, David Varnes, Robert Schuster, and Arvid Johnson. The application of slope stability studies to land management questions has, in turn, been advanced by Earl Brabb, Douglas Swanston, Fred Swanson, Raymond Rice, Robert Ziemer and Donald Coates. These people and many others have either directly or indirectly influenced my approach to the study of landslides.

The perspectives of this book, to some extent, also reflect my professional activities and research experience. Inevitably, as an academic, I have tried to abstract the underlying principles of various approaches to the landslide problem. It is, thus, not a book on techniques or methods of analysis, nor is it a compendium of detailed case studies. However, my outlook has also been shaped by involvement in occasional consulting work and legal hearings which have served to make me acutely aware of the need to isolate and communicate precisely cause and effect aspects of the slope stability question. This objective therefore influences the presentation of much of the material in the book.

An important stimulus for writing Landslides: Causes, Consequences and Environment arose out of my work with the Task Force on Natural Hazards of the New Zealand Commission for UNESCO. The Task Force activities brought me into contact with a wide range of professional people who were concerned in one way or another with landslides and similar natural hazards. Besides earth scientists and engineers these include: land managers, planners, meteorologists, foresters, public policy advisors, lawyers, land valuers, and insurance specialists. I realised that such a book could be of use to these people, as well as to students of earth science.

The opportunity to do fieldwork and to view the landslide problem first-hand in a number of different countries has also broadened my perspectives on the subject. These countries include: Canada, New Zealand, United States of America, Peru, Ecuador, Fiji, New Caledonia, Australia, England, France, Spain and the Soviet Union. It would not have been possible to gain this experience without the financial and logistic support of a number of organisations: The Fulbright Foundation; The Forestry Sciences Laboratory, Corvallis, Oregon (United States Department of Agriculture, Forest Service); UNESCO; Victoria University of Wellington; New Zealand University Grants Committee; Trent University, Ontario; University of Alberta; Soil Conservation and Rivers Control Council, Ministry of Works and Development (New Zealand); University of Otago; Office de la Recherche Scientifique et Technique Outre-Mer; Department of External Affairs of French Government; Applied Geology Associates Limited, New Zealand; United States Geological Survey, Menlo Park; Instituto de Geologia y Mineria, Peru.

I would also like to express my gratitude to some stimulating colleagues for helping to keep mind and body together in the sometimes rigorous conditions of the field. In this respect thanks are owed to Peter Adams, Bob Eyles, Ian Grant, Jacques Iltis, George Lienkaemper, Jack McConchie, Sally Marx, Rob Owen, Fred Swanson, Ralph Wheeler, and Marco Zapata.

For assistance in the production of this book I am grateful to Robin Mita, cartographer of the Geography Department, Jean Benfield, and John Casey and his staff of the photographic unit, Victoria University of Wellington. In particular I am indebted to the Wellington office of Applied Geology Associates Limited and its Director, Martin Ward, for providing word processor and editorial assistance. Thanks are owed specially to Sally Marx and Rob Owen for editorial assistance and to Vanessa Allen for her expertise and patience with the word processor and me.

For understanding the task, thanks are owed to Christina, Lara and Sarah.

M J Crozier
Research School of Earth Sciences
Victoria University of Wellington

ACKNOWLEDGEMENTS

Acknowledgements are made of the following sources:

Aerial Surveys Ltd, Nelson, New Zealand: (Figure 4.10a).
American Society of Civil Engineers: Holtz, W G; Gibbs, H J (1956). Transactions of the American Society of Civil Engineers 121: 641-77. (Table 3.2).
Catena Verlag: Young, A R M (1978). Catena 5: 95-107. (Table 5.7).
Department of Geography, Victoria University of Wellington: Crozier, M J; Howarth, R; Grant, I J (1981). Pacific Viewpoint 22(1): 69-88. (Figure 4.18).
Department of Lands and Survey, New Zealand: (Figure 4.19).
Department of Scientific and Industrial Research, New Zealand: Crozier, M J; Eyles, R J; Marx, S L; McConchie, J A; Owen, R C (1980). New Zealand Journal of Geology and Geophysics 23: 575-86. (Figure 4.22 and Table 4.8).
Eyles, Dr R J: (Figure 5.4b).
Gebruder Borntraeger: Crozier, M J (1973). Zeitschrift fur Geomorphologie 17(1): 78-101. (Figure 2.2 and Tables 2.3; 2.5). Iida, T; Okunishi, K (1983). Zeitschrift fur Geomorphologie, Supplement 46: 67-77. (Figures 4.7; 4.9). Selby, M J (1980). Zeitschrift fur Geomorphologie 24(1): 31-51. (Table 3.8).
Hawkes Bay Catchment Board, Napier, New Zealand: (Figure 4.10b).
Hutchinson, Dr J N: (Tables 2.2; 2.7).
International Society for Rock Mechanics: Hoeke, E; Londe, P (1974). Proceedings of 3rd Congress, Denver: 613-752. (Figure 3.13).
McKenzie, Prof D W: (Figures 3.12; 4.12; 4.13; 4.14).
Ministry of Works and Development, New Zealand, Photo Library: D Bircham. (Figure 6.4).
National Academy of Sciences, Transportation Research Board, Washington DC: Varnes, D J (1978). Landslides: Analysis and Control. Special Report 176. Edited by R L Schuster and R J Krizek. (Tables 2.8).
National Water and Soil Conservation Organisation, Ministry of

Works and Development, New Zealand: Grant, P J (1983). Soil
Conservation Centre, Aokautere, Publication 5. (Figures 4.16;
4.17).
New Zealand Alpine Club: Chinn, T J (1979). New Zealand
Alpine Journal 32: 35-7. (Figure 4.1).
New Zealand Geographical Society: Eyles, R J; Crozier, M J;
Wheeler, R H (1978). New Zealand Geographer 34(2): 58-74.
(Figures 5.1; 5.5; 5.6; 5.7; 5.8 and Tables 5.3; 5.5; 5.6).
New Zealand Geographical Society, Wellington Branch:
Eyles, R J; Eyles, G O (1982). Proceedings of 11th Geography
Conference, Wellington: 118-22. (Figures 4.3; 5.3; 5.4a).
New Zealand Hydrological Society: Owen, R C (1981). Journal
of Hydrology (New Zealand) 20(1): 17-26. (Figure 4.23).
New Zealand Institution of Engineers: Riddolls, B W;
Perrin, N D (1975). New Zealand Engineering 30(8): 221-5.
(Figure 3.18). Crozier, M J; Eyles, R J (1980). Proceedings
of 3rd Australia-New Zealand Conference on Geomechanics,
Wellington, 2: 2.47-2.51. (Figures 5.10; 5.11; 5.12; 5.13).
Oxford University Press: Selby, M J (1982). Hillslope
Materials and Process. (Table 3.6).
Otago Daily Times, Dunedin, New Zealand: (Figure 6.2).
Pitman Publishing Inc: Lee, I K; White, W; Ingles, O G
(1983). Geotechnical Engineering. (Table 3.2).
Royal Geographical Society Australia, Queensland Inc: East, T
J (1978). Queensland Geographical Journal, 3rd Series (4):
37-67. (Figure 2.4).
Stephens, P R: (Figure 6.5).
United Nations Disaster Relief Organisation, Geneva (1979):
Disaster Prevention and Mitigation: Volume 7 Economic
Aspects. (Table 6.1).
United States Department of Agriculture, Forest Service,
Pacific Northwest Forest and Range Experiment Station:
Dietrich, W E; Dunne, T; Humphrey, N F; Reid, L M (1982). In:
Sediment Budgets and Routing in Forested Drainage Basins.
Edited by F J Swanson, R J Janda, T Dunne, and D N Swanston.
General Technical Report PNW-141: 5-25. (Figures 4.11; 4.15).
John Wiley and Sons Ltd: Crozier, M J (1984). In: Slope
Instability. Edited by D Brunsden and D B Prior: 103-42.
(Figure 3.5, Table 6.3 and Appendix 2).
Wellington Regional Water Board, New Zealand: Bishop, R J
(1977). Report on Storm of 20 December, 1976. (Figure 5.2).

THE APPROACH   **1**

Landslides do not occur in isolation; they are a product of their environment and in turn they influence its condition. That is what this book is about. It demonstrates that the internal workings of the slope which lead to failure can be linked to past and present environmental factors. The human factor is an important part of that environment not only because of man's ability to manipulate it but also because of his own vulnerability.

Much of the literature on landslides has been event-specific, often consisting of case studies for scientific, engineering or forensic purposes. Understandably, engineers and engineering geologists who have contributed so much to this literature are required to solve one problem at a time. Indeed, their case studies have been instrumental in stimulating and complementing systematic research into the geomechanics and physics of failure mechanisms and controlling forces. As a result, what goes on during a landslide is reasonably well understood, in strict physical terms. What is less well understood, is why those forces should combine to reach a critical state at one specific time and in one specific place. The solution of this problem requires not only a knowledge of the functional link between environmental and slope conditions but equally an understanding of the dynamics of the environmental system.

An attempt to identify critical environmental conditions solely from slope physics can soon confront the student with a number of paradoxical questions. It has been shown, for example, that tree roots impart a strength to the slope that can be quantified and treated as an increment of slope resistance (O'Loughlin and Ziemer 1982). Why, then, did So (1971) find that forested areas were more susceptible to landslides than were other areas? Using another example, many studies (for example, O'Byrne 1967) have shown that slope resistance is related to rock type. Why then could Rhodes (1968) make the observation that granitoid rocks in New Guinea gave rise to a greater density of landslides than did any other rock type while Radbruch and Crowther (1973) found that the same sorts of rock in California were amongst the most stable? Why, in some hazard classification schemes (Cooke and Doornkamp 1974) are steep slopes taken to indicate states of potential instability when, in other instances (Young 1961), such slopes are considered to denote the presence of strong rocks?

In yet another example of a paradoxical situation, how should deeply weathered slopes be classified on a map depicting landslide susceptibility? Slopes with a thick cover of weathering products apparently reflect surfaces which have been stable for a long period, yet there are many instances (Durgin 1977) where such slopes are highly susceptible to landslides. Certain sites which have been subject to shallow landslides can be considered to be relatively immune from further failure (Crozier et al. 1982) yet Dietrich et al. (1982) note that such sites may give rise to recurring landslide activity. Why does the imposition of weight lower stability on some sites and increase it at others?

These apparent contradictions and questions are typical of those which may confront the student who is trying to come to terms with the diverse literature on landslides. The broad, yet systematic, approach taken in this book, however, is aimed at providing a framework within which such contradictions should not arise. Whereas geomechanics and allied disciplines provide an understanding of landslides from an 'internal' perspective, this book provides, in addition, an 'external' perspective. Nevertheless, fundamental geomechanical principles are essential and in the first part of the book they are systematically discussed and subsequently linked with the complex and multivariate environmental framework within which they operate. This is done by emphasising cause and effect aspects of the slope system and discussing the changes to critical conditions in space and time brought about by historic and geographic changes within the environment.

Understanding the causes of landslides imposes a responsibility on the community to mitigate the impact of such phenomena. The book therefore concludes by looking at the social and environmental costs of landslides and the kind of information base and decisions that are required to cope adequately with the problem.

Slope stability is a specialised topic within environmental science, indeed even within earth science and, as such, it is best understood with a general grounding in those subjects. However, this book has not been written for the landslide expert but for those who come into contact with the slope stability question in the course of their work or studies.

CLASSIFICATION OF SLOPE MOVEMENT

This chapter is not intended to provide a detailed description of all the different types of slope movement, as descriptive classifications can be readily found in a number of publications: Sharpe 1938; Ward 1945; Campbell 1951; Varnes 1958; Yatsu 1966; Hutchinson 1968; Skempton and Hutchinson 1969; Zaruba and Mencl 1969; Crozier 1973; Northey, Hawley and Barker 1974; Hutchinson 1977; Varnes 1978. The proliferation of classifications in recent years however, has unintentionally defeated one of the principal purposes of the exercise; that is, the provision of clear and unambiguous terminology. This has arisen because individual classifications have often been developed for specific purposes or from observations in specific regions and their subsequent application has not always taken this into account. A classification which is highly appropriate locally may be unsatisfactory in another area or ambiguous when correlated with other classifications. The purpose of this chapter therefore is to provide guidance in selecting and applying the most appropriate classification. However, before this can be done, the bases of commonly used classifications must be identified and their terminology evaluated.

2.1   WHAT MAKES A GOOD CLASSIFICATION?

Classification is sometimes considered as merely a first step to scientific investigation, or as an inferior form of activity best left to academics who have nothing better to do. In reality, however, classification is a powerful process in the transfer of ideas: classifications institutionalise concepts and are therefore both valuable and dangerous.

A classification is designed to reduce a multitude of different but related phenomena to a few easily recognised and meaningful groups on the basis of common attributes. The process involved is neither pedestrian nor simply a prelude to investigation; indeed it is akin to model building or the development of formulae, wherein the maximum or required degree of explanation is provided by the minimum number of parameters. Classifications are also constantly evolving with the acquisition of new information and a good one will accommodate new findings with the minimum of structural change.

A good classification will also specify its classificatory parameters in unambiguous, universal terms to allow for standardised application and reproducibility of results. Even more fundamental, however, is the choice of those parameters. They must be distinguishable by the measuring techniques available, and they must reflect either natural boundary conditions for slope processes or features important to the end use. In addition, they must produce groups of phenomena which are not only internally consistent but which are also meaningful in terms of the purpose of the classification.

If the classification adheres to these tenets it should be successful. It should allow us to talk economically about the important features and to ensure that everybody knows at least 'what' they are talking about. It is thus important for the practitioner to understand the purpose behind each classification in order to choose the right one for the right job.

2.2   SOME FUNDAMENTAL DIFFERENCES

Classifications of slope movement often have different titles; in some cases these reflect only a semantic difference but in others they indicate a variation in the scope of treatment. Some headings commonly encountered are: slope failure (Ward 1945), mass wasting (Yatsu 1966), mass movement (Hutchinson 1968), landslides (Varnes 1958) and slope movement (Varnes 1978). To understand the differences implied, it is necessary to clarify terminology and to place slope movements in the context of other earth-forming processes.

2.2.1   TERMINOLOGY

In strict geological terms, slope movements constitute a part of the denudational process which itself is one of four major groupings within the rock cycle (**Figure 2.1a**). Denudation is an old and broad concept referring to wearing away of the landmass through geological time.

In the writings of 19th century geologists denudation was clearly conceived as the net effect of all those processes, either sub-aerial or marine, which removed the landmass and contributed material to the sedimentary process (Lyell 1853). Davis (1896) noted that contemporary geologists equated the process with 'base-levelling' but six years later (Davis 1902) he was advocating a more restricted use for the term. He wanted the use of 'denudation' to be confined to active removal of regolith or waste rock from slopes, a process that

a. ROCK CYCLE

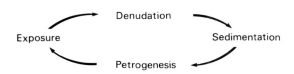

b. DENUDATIONAL PROCESSES

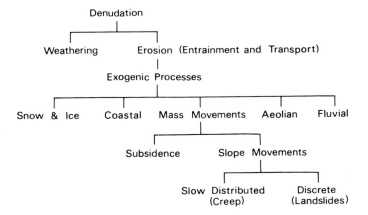

c. NORMAL EROSION

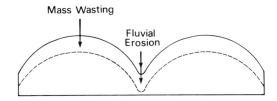

Figure 2.1   Classification of major geomorphic processes

would occur in the sort of steeplands depicted by the early stages of his 'geographical cycle' of landform evolution. He insisted that the term 'degradation' should be employed for those less active processes that gradually reduced the gentler, waste-covered slopes during the later stages of the cycle. Thus in contrast to its broader use, Davis excluded river action from the term denudation and although the precise distinction between denudation and degradation has largely been forgotten, some modern geomorphologists (Young 1972) still persist with Davis' definition by excluding the work of streams and other linear processes from the concept of denudation. They refer to the removal of material by linear processes as erosion which is an unfortunate distinction because much of the mass removal of material from slopes is also widely known as erosion. It is much less confusing to adhere to the original and more generally understood usage of the terms indicated in **Figure 2.1b.**

Denudation then ultimately involves mechanisms which weaken the rock (weathering) and pick it up and carry it away (erosion). Five groups of processes (Figure 2.1b) are recognised as being capable of erosion and, because the intensity of their activity is controlled largely by forces external to the earth, they are termed exogenic (exogenetic) processes. Mass movement is distinguished from the others by being the outward or downward gravitational movement of earth material without the aid of running water as a transportational agent. This is a very precise and useful definition as it does not deny the importance of water in either its solid or liquid state as a destabilising factor nor does it exclude subsidence and other movement on flat ground.

What then is mass wasting? In usage, it often appears to be synonymous with mass movement but it is really a broader geomorphic concept commonly used in conjunction with the erosion cycle to refer to the mass reduction of the interfluves as opposed to the degradation by streams **(Figure 2.1c).** In effect it must include the action of all non-linear erosional processes working on the slopes between streams.

Slope instability is another general term which refers to the predisposition of a slope to mass movement. The condition may be recognised by analysis of stress within the slope, by various slope characteristics or by analysis of the historical record of slope development. The issues involved in the definition of slope instability are discussed in Chapter three.

The term landslide has enjoyed the most common and universally appreciated currency as the collective term for most slope movements of the mass movement type and will probably retain this status for some time into the future. It has generally been seen as a category of mass movement excluding creep and subsidence. However, here the agreement ends, as there is a variety of opinion as to whether it should encompass 'falls', 'flows' and certain other movements including snow- and ice-related activity. Skempton and Hutchinson (1969) decided that: 'The generic term, landslide, embraces those down-slope movements of soil or rock masses as a result of shear failure at the boundaries of the moving mass'. However, even defined in this way, the term could include many movements where nearly all the displacement occurs by flow rather than slide. Thus 'landslide' has on occasions been considered unsuitable as a broad collective term (Crozier 1973; Varnes 1978) because the active part of the word denotes sliding. Varnes (1978) has advocated the term slope movements for mass movement restricted to slopes and as this appears to be a suitably neutral, all-embracing term, it is used in this sense throughout this book.

For practical purposes, however, it is still useful to distinguish between two broad categories of slope movements: 'discrete' and 'slow and distributed' (Figure 2.1b). Discrete slope movements correspond more or less to Skempton and Hutchinson's (1969) definition of 'landslides' and are recognisable as areally discrete and often rapid movements whereas slow and distributed movements involve ill-defined extensive areas of slow movement including, for example, 'shallow creep'.

The remaining broad collective term widely used is slope failure. Although it is free from mechanistic connotations it is more accurately used, as Terzaghi (1950) advocated, for slope movements on engineered slopes. The term, however, is sometimes used to refer to the process of rupture or shearing in material rather than to a particular ground feature.

Mass movement grades imperceptibly into the fluvial process with increasing water content. In practice it can be difficult to make the distinction between some types of fluid mass movement and certain forms of fluvial activity such as intermittent, highly sediment-charged, torrents. However the following field criteria can be used to make the distinction between fluid mass movements and fluvial processes:

1. mass movement deposits usually come to rest with convex-upward surfaces or lobes whereas fluvial deposits have more concave-upward longitudinal surface profiles;
2. because of the mechanics of fluid mass movement, large boulders can be transported on the surface or at the front of flows and may subsequently be left stranded on levees or as perched deposits. Fluvial processes in contrast transport the coarsest debris as bedload;
3. small cavities or 'air pockets' are sometimes formed between blocks of mass movement debris. Cavities are not generally found in fluvial deposits;
4. although rock particles may show little sorting in mass movement and fluvial torrent deposits, small woody debris tends to be floated into segregation lenses by the fluvial process.

7

2.2.2   AIMS AND CRITERIA

The specific criteria used to categorise slope movements depend on the particular aim of the exercise which in turn relates to the nature of the task in front of the investigator. Some aims may be associated with a passive task: the classification of land for potential use (for recreation, primary production, industry, housing and/or communications) or the assessment of geomorphic activity. Others may be associated with an active task such as stability analysis for control, prevention or avoidance.

Varnes (1978) provides a summary of the criteria used in classifications which, expanded a little, includes:

1. type of movement: recognised as falls, topples, slides, lateral spreads, flows;
2. kind of material: rock (incompetent, competent), engineering soil (regolith and sediments), true (pedological) soils, regolith (mantle), debris (predominantly coarse engineering soil), earth (predominantly fine engineering soil), clay (particles finer than 2 microns), mud (a mixture of clay and silt), material defined by mineralogy;
3. quality of material: degree and orientation of structural discontinuities (joints, fissures, faults and bedding), particle characteristics, strength characteristics (ranging from residual to peak shear strengths), degree of sensitivity and thixotropy;
4. morphometric characteristics: depth, dimension ratios (especially depth/length) and geometry of failure surface, transport path and depositional form;
5. degree of displacement of material from its in situ position;
6. degree of disruption of the displaced mass;
7. orientation of slide geometry to geologic structure and land form;
8. age, state of activity and history of movement;
9. geographic location or climate zone of type examples;
10. geologic setting;
11. rate of movement;
12. water, air and ice content: their effect on cohesion, internal friction and pressures exerted on failure surface;
13. degree of potential hazard;
14. causes.

The choice of criteria and the degree to which they can or should be assessed in the application of a classification also depends very much on the cost/benefit aspects of the exercise, the scale of the investigation, as well as the technological and financial resources available.

Three examples will serve to illustrate some of these points. The first of those chosen represents an attempt to classify slope movements in a way which is relevant to soil conservation objectives. The second example is designed to highlight types of movement which reflect significant geotechnical differences and the third concentrates on morphometric characteristics.

Example 1: Mass Movement Erosion (Campbell 1951)

This first example (Table 2.1) represents an attempt by Campbell (1951) to provide a scheme whereby an initial inventory of a country's erosion problems could be achieved - the scheme is still in use in New Zealand.

Table 2.1   Mass Movements and Soil Erosion (Campbell 1951)

Simple types	Compound types

1. **Creep erosion** -
    Soil creep
    Scree creep
    Solifluction
	Sheet and creep
	Terracettes

2. **Slip erosion** -
    Soil slip
    Earth slip
    Slump
    Subsidence
	Slip and earthflow
	Slump and earthflow

3. **Flow erosion** -
    Earthflow
    Debris avalanche
    Mudflow
	Creeping earthflow

Associated erosion

4. **Sheet erosion**	Sheet and wind erosion
5. **Gully erosion**	Gullied earthflow Gullied slip Tunnel-gully

6. **River erosion** -
    Bank erosion
    Sedimentation	

7. **Wind erosion**	Wind and sheet erosion

The scheme is a direct descendant of Sharpe's (1938) mass movement classification but has been expanded to include aspects of all common geomorphic processes that might be detrimental to the soil resource. The scope and emphasis of the classification is easily appreciated when it is realised that Campbell was, at the time, working for the New Zealand Soil Conservation and Rivers Control Council. This body was responsible for administering the Soil Conservation and Rivers Control Act 1941 which directed it to 'prevent flood damage, mitigate soil erosion and promote soil conservation'. One of the Council's primary functions was to carry out an erosion survey in the early 1950's to assess, in broad terms, the nature and severity of the threat to the soil resource.

The classification gives little attention to parameters such as failure geometry, strength properties or type of material that might assist in analytical work or provide clues to the causes. It is based mainly on existing and secondary characteristics rather than the mechanisms of initial movement, as control and rehabilitation were more immediate concerns than were prediction and prevention. For example, the distinction made by Campbell between 'soil slip' and 'earth slip' depends on the degree to which the pedological soil has been removed and thus, by implication, the ease with which pasture cover can be re-established.

Although the classification served its purpose in the early days of New Zealand's soil conservation history, it is severely limited in meeting the kind of modern demands placed on slope movement classifications. It would be a mistake, for instance, to make even broad land use decisions, particularly for urban developments, from land inventory data compiled solely on the basis of this sort of classification. The classification simply does not direct sufficient attention to bedrock and deep-seated movements nor does it distinguish soil movements peculiar to unstable materials such as certain ash soils and montmorillonite-rich rocks which are encountered in New Zealand.

Despite this, Campbell's classification continues to be used by some New Zealand government agencies in essentially its original form. It still has some value as a broad measure of soil erosion problems but even in this role the scheme could be improved. For example, four attributes could easily be recorded in the course of classifying each slope movement, thus greatly enhancing the value of the recorded data and making the application of the scheme more appropriate to the kind of problems currently being faced by soil conservators. The attributes are:

1. whether debris from slope movements identified was contributing directly to stream load. This would help to indicate likely downstream problems from channel aggradation;

2. the calibre of debris, thus indicating whether

streams were to be supplied with suspensed or bed load material;

3. the position of the slope movement in the landscape. This would give some indication as to the likely success of revegetation because of microclimatic and soil moisture considerations;

4. the depth of movement and degree of activity which would indicate the likelihood of tree roots becoming established through the displaced mass.

Example 2: Geotechnical Classification (Hutchinson 1978)

The second example, chosen to indicate how an end-use dictates the nature of the classification, is provided by Hutchinson's (1978) geotechnical classification (**Table 2.2**). The scope is totally different from that of the previous example because there is a very narrow and immediate objective; that is, to enable successful stability analysis to be achieved.

Stability analysis (see Section 3.3.1 RESISTANCE AND SHEAR STRESS) is carried out by determining the difference between the amount of shear (or rupturing) stress being exerted within the slope and the amount of resistance that can be mobilised against that stress by the material involved. Even in ground that shows no sign of movement, if this difference is very small the situation is just marginally stable, as only a small rise in shear stress or a small drop in resistance is required to cause a slope movement.

Stability analysis is also often applied to existing landslides to gain some idea of which factor or factors could have changed sufficiently to have upset the balance of forces that existed prior to movement. In either of these cases, detailed information is required on the shape and weight of the mass involved, much of which will be only indirectly indicated by 'general' slope movement classifications. Much of the essential information then needs to come from accurate surveying and sampling and laboratory testing of the material.

The strength parameters of material (cohesion and 'internal friction') which make up an important part of a slope's resistance to movement has to be determined with the aid of testing equipment. However, different strength values can be obtained from the same sample depending on how the test is conducted and how the results are interpreted. This makes it extremely important to carry out tests in a way which will provide the sort of strength values pertaining not only to the material on the shear plane (the place where rupture occurs) but also to the manner and speed at which the shear stress is likely to be exerted within the ground.

Some landslides occur as a result of a sudden increase in shear stress which is often caused by steepening and increasing the height of the slope through excavation. One explanation of what actually happens in this case is that

11

Table 2.2   Geotechnical Classification of Landslides
After A W Skempton and J N Hutchinson
(Hutchinson 1978)

Soil fabric conditions (affecting cohesion and internal friction)	Pore fluid pressure conditions on slope surfaces (affecting porewater pressure)
1. First-time slides in previously unsheared ground: soil fabric tends to be random (or oriented as a result of depositional history) and shear strength parameters are at peak or between peak and residual values. 2. Slides on pre-existing shears associated with: (a) previous landslides (b) colluvium (c) periglacial solifluction (d) other freeze-thaw processes (e) tectonics (f) lateral expansion In these cases the soil fabric surface is highly oriented in the slip direction and shear strength parameters are at or about residual values.	A. Short-term (undrained) - no equalisation or excess porewater pressure set by changes in total stress. B. Intermediate - partial equalisation of excess porewater pressures. C. Long-term (drained) - complete equalisation of excesss porewater pressures to steady seepage values. Note that combinations of A, B and C can occur at different times in the same landslide; for example, a particularly dangerous type of slide is that in which long-term, steady seepage conditions (C) exist up to failure but during failure, undrained conditions (A) apply; that is, a drained/undrained failure.

during excavation, as the weight of overburden material is removed, the exposed material is unloaded and expands thus reducing the pressure of water contained in spaces (pore pressure) within the material. The reduction in pore pressure tends to increase the resistance of the slope, but if this is insufficient to counteract the incurred shear stress imposed by the new form of the slope, a landslide will occur - known as a 'short-term' landslide. The failure happens so quickly that, given a fine-grained material, water has little chance to drain from the site.

If one is anticipating or analysing such a failure then strength testing in the laboratory or the field must try to replicate these conditions by using what is referred to as total stress and undrained conditions within the sample (explanation of these terms and appropriate testing procedures are discussed by Petley 1984).

'Long-term' landslides on the other hand are not triggered by rapid increase in shear stress, but either by a slow, progressive increase in shear stress or commonly by an increase in porewater pressure which has the effect of reducing resistance within the slope. In this case the effect of porewater pressure generated during the test must be eliminated, or measured and subtracted from total stress in order to obtain effective strength values. Consequently either a drained test is performed or, if appropriate, an undrained test in which porewater pressure is measured.

Because the kind of rock or soil conditions and the pore fluid pressures determine the manner of laboratory test required to produce accurate results, it is obvious that these geotechnical conditions should be taken into account in classifying slope movements. Hutchinson, in his geotechnical classification, has used two parameters which give a useful guide to the manner in which testing should be carried out and the values to be considered in stability analysis. One is the history of movement to which the slope material has been subject, and the other is the time between the formation of the slope and the ensuing landslide.

Example 3: Morphometric Classification (Crozier 1973)

Crozier (1973) carried out a study of various form characteristics of shallow slope movements occurring in unconsolidated soils and slope deposits in the South Island of New Zealand. Most of these landslides exhibited translational flow movement and the aim of the study was to characterise the degree and type of this movement by using simple morphometric indices. Arguing that in a given material the distance of flow would be a function of slope angle, volume of material and water content, Crozier developed a method of isolating the degree of flow deformation of the displaced mass that could be attributed to water content. The method thus offers a way of reconstructing the probable consistency of the material at the time of the event. Whereas most other factors promoting flowage can be measured after the event, water content at failure is a temporary condition which can only be measured by sampling during the actual event - something which is rarely, if ever, achieved. In addition, Crozier (1973) and others (Cooke and Doornkamp 1974) have found that these morphometric indices can also be used to distinguish various classes of movement recognised by other workers.

The seven indices used are listed in **Table 2.3** and the measurements required to use the method are indicated in **Figures 2.2 and 2.3**. The depth index has been used by a number of workers to provide a relative measure of surficiality. The dilation and tenuity indices measure the

Classification

Table 2.3   **Morphometric Indices** (Crozier 1973)
(see Figure 2.2 for measurements)

Index	Calculation
Depth	$\dfrac{D}{L} \times 100\%$
Dilation	$\dfrac{Wx}{Wc}$
Tenuity	$\dfrac{Lm}{Lc}$
Flowage	$\left\| \dfrac{Wx}{Wc} - 1 \right\| \dfrac{Lm}{Lc} \times 100\%$
Viscous Flow	$\dfrac{Lf}{Dc}$
Displacement	$\dfrac{Lr}{Lc}$
Fluidity (water content)	Ranked residuals from regression of flowage on slope (see Figure 2.3)

degree of spread in the lateral and downslope axial directions respectively and the flowage index incorporates both these measures to provide an indication of bi-axial spread. Any of these three spread indices can be regressed against slope, and, given constant material properties, the resultant residual can be ranked to provide a measure of fluidity (Figure 2.3), which is primarily a function of water content.

The displacement index measures the degree to which the displaced mass has evacuated the surface of rupture. In a sense it indicates the stage of development of the landslide and provides a rough guide to the residual stability of the slope. Partially evacuated surfaces with low displacement values allow ponding of water upslope of the displaced mass, which itself rests upon sheared material reduced by movement to its residual strength. Such conditions usually permit re-activation to occur with conditions of lower magnitude than the original triggering event.

Crozier (1973) demonstrated the way in which these indices might be used to classify various types of landslide. Five process groups (**Table 2.4**) were established for 66 landslides using the descriptive criteria of Varnes (1958). Measures of central tendency were then found for the

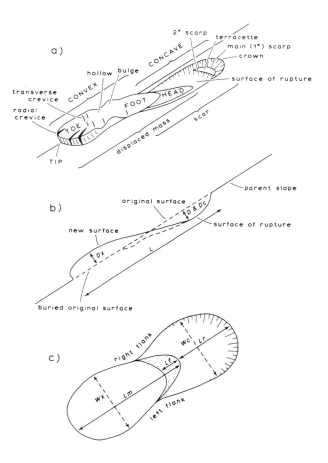

Figure 2.2   **Terminology used in the morphometric
classification** (Crozier 1973)

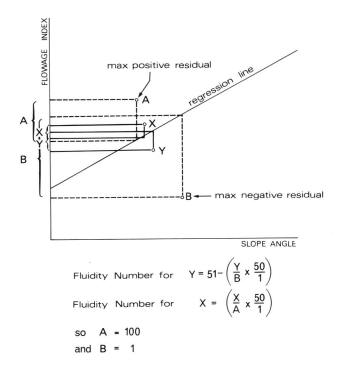

$$\text{Fluidity Number for} \quad Y = 51 - \left(\frac{Y}{B} \times \frac{50}{1} \right)$$

$$\text{Fluidity Number for} \quad X = \left(\frac{X}{A} \times \frac{50}{1} \right)$$

so A = 100
and B = 1

Figure 2.3   Graphic determination of fluidity index

Table 2.4   Process Groups used in Morphometric Analysis (Crozier 1973)

Process Group	Class of Movement
Fluid-Flow	Mudflows, debris flows, debris avalanches
Viscous-Flow	Earthflows, bouldery earthflows
Slide-Flow	Slump/flow
Planar Slide	Turf glide, debris slides, rock slides
Rotational Slide	Earth and rock slumps

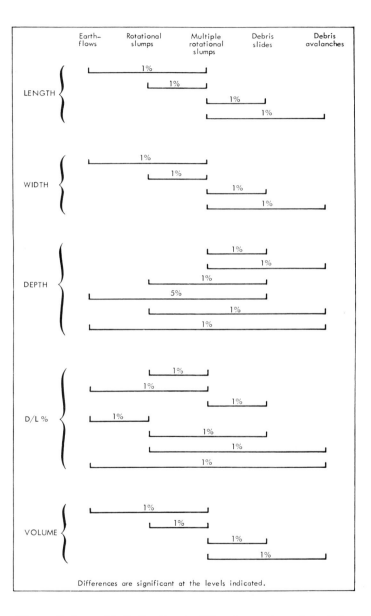

Figure 2.4 **Statistically significant morphometric difference between mass movement types** (East 1978)

Table 2.5   Statistical Summary of the Morphometric
             Indices of each Process Group (Crozier 1973)

Index		Process Group				
		RS	PS	SF	VF	FF
Depth	m	24.23	7.66	4.98	3.34	1.47
	s	19.28	5.92	1.95	1.39	1.00
	m + s	43.51	13.58	6.93	4.73	2.47
	m − s	4.95	1.74	3.03	1.95	0.47
Dilation	m	0.99	0.95	0.94	1.09	0.89
	s	0.10	0.09	0.19	0.37	0.41
	m + s	1.09	1.04	1.13	1.46	1.30
	m − s	0.89	0.86	0.75	0.72	0.48
Flowage	m			16.01	5.39	12.14
	s			5.88	4.57	10.59
	m + s			21.89	9.96	22.73
	m − s			10.13	0.82	1.55
Displacement	m	68.10	79.87	56.89	29.28	59.06
	s	12.03	6.00	20.47	30.30	30.87
	m + s	80.13	85.87	77.36	59.58	89.93
	m − s	56.07	73.87	36.42	0	28.19
Viscous-Flow	m				3.66	
	s				2.41	
	m + s				6.07	
	m − s				1.25	
Tenuity	m	13.19	1.17	3.07	1.71	3.33
	s	24.15	0.04	0.20	0.71	1.94
	m + s	37.34	0.21	3.27	2.42	5.27
	m − s	0	1.13	2.87	1.00	1.39

m = mean   SF = Slide-Flow
s = one standard deviation   VF = Viscous-Flow
RS = Rotational Slide   FF = Fluid-Flow
PS = Planar Slide

indices of each group **(Table 2.5).** The extent to which these indices can successfully discriminate between the process groups by two factor plots is illustrated by Cooke and Doornkamp (1974) and summarised in **Table 2.6.** The depth index by itself, or in conjunction with other indices, is clearly the most versatile in distinguishing process groups.

In a study of the morphometry of 99 landslides in south-east Queensland, Australia, East (1978) also found that the depth index was the most successful surrogate measure of process. **Figure 2.4** shows that the depth index is a successful discriminating parameter in 7 out of the 10 possible combinations, only failing to distinguish the following groups: earthflows/debris slides; multiple rotational slumps/debris avalanches; debris avalanches/debris slides.

Two other notable studies have examined the application of morphometry to landslide systems (Brunsden 1973) and to landslide classification (Blong 1973). Both conclude that

Table 2.6   Indices Most Suitable for Discriminating Between Process Groups
(after Cooke and Doornkamp 1974)

	FF	VF	SF	PS	RS
FF	X	depth* v flow	depth	depth v tenuity	depth
VF		X	flow	depth v displace	depth
SF			X	depth* v displace	depth* v tenuity
PS				X	depth* v tenuity
RS					X

* Incomplete discrimination at $\pm$ one standard
  deviation

FF = Fluid-Flow   PS = Planar Slide
VF = Viscous-Flow   RS = Rotational Slide
SF = Slide-Flow

morphometric parameters are useful as supplementary discrip-
tors of process. However, Blong's effort to match morpho-
metrically based grouping with Varnes' (1968) debris slide,
debris flow and debris avalanche groups showed that the
morphometric-based classification and the process-based
classification were incompatible.

2.3   GENERAL CLASSIFICATIONS

The three classifications discussed above represent
widely divergent approaches to the way in which slope movements may be characterised and classified. Each has been
designed for a specific purpose and the scope of their application is thus largely limited to that objective. However,
where an initial assessment is being made of slope movements
or where a study has multiple objectives, a more general
classification is often required.

The two generalised classifications most likely to be
encountered in the English speaking world are authored respectively by J N Hutchinson (1968, 1969 (with A W Skempton),
1977) of Imperial College, London and D J Varnes (1958, 1978)
of the United States Geological Survey, Denver. These two men
have brought a high level of expertise and wide experience to
bear in the compilation of their classifications (**Tables 2.7
and 2.8**) and their full work should be consulted before making
any serious attempt to classify slope movements.

Both authors use 'type of movement' to establish the
principal groups. The only exception is found where
Hutchinson delineates one of his five major groups
(freeze-thaw phenomena) on the basis of the geo-climatic location or climatic function. It would be more consistent, if
not particularly practical, to distribute most of the freeze-
thaw phenomena among the other principal movement categories.
The argument for singling out this group is that the phenomena
occur in particular environmental conditions not experienced
by other slope movements and that they possess distinctive
rates and periodicity of movement.

The major distinction between the two classifications is
the difference accorded to the status of flow movement.
Hutchinson, like his early British predecessor Ward (1945),
does not recognise flow as a primary mechanism of failure -
Varnes, on the other hand, like his American predecessor
Sharpe (1938), does. Even in the one publication (Skempton
and Hutchinson 1969) where Hutchinson treats flow as having
status similar to other movement groups (at least for
discussion) the point is made that: '....the distinction between slides and flows drawn by Sharpe (1938), which is central
to his classification of mass-movements, is not generally true
(Hutchinson 1965)'. While it is evident that many mudflows
and earthflows exhibit varying amounts of sliding at the con-

Table 2.7   Mass Movements on Slopes (Hutchinson 1977)

A. ELASTIC REBOUND

   1. Inward movements of valley sides

   2. Vertical rebound of valley floor
     (a) Valley anticlines
     (b) Raised valley rims and upward flexure valley
        walls

B. CREEP

Generally not leading to a landslide
   1. Shallow, predominantly seasonal creep; Mantle creep
     (a) Rock creep (terminal curvature)
     (b) Soil creep
     (c) Talus creep

   2. Deep-seated continuous creep; Mass creep

Generally leading to a landslide
   3. Pre-failure creep, progressive creep (primary,
     secondary, tertiary)

Post slide
   4. Post-failure creep

C. LANDSLIDES

   1. Rotational slips/slides (each unit approximately
     circular)
     (a) Single rotational slips (single event)
     (b) Successive rotational slips (usually retro-
        gressive, occasionally progressive)
     (c) Multiple rotational slips (all retrogressive)
        (i) in stiff, fissured clays
        (ii) in soft, extra-sensitive (quick) clays;
           clay flows

   2. Compound slides (markedly non-circular)
     (a) Graben slides (single event)
     (b) Progressive non-circular slides

(continued...)

Table 2.7 (continued)

---

3. Translational slides (usually single event or retro-
    gressive)
    (a) Rock slides
        (i) Planar slides; block glides (two-dimens-
            ional)
        (ii) Wedge failures (three-dimensional)
    (b) Slab, or flake slides
    (c) Spreading failures (all retrogressive)
    (d) Debris (detritus) slides
    (e) Mudslides (generally retrogressive)
    (f) Bog slides, bog flows, bog bursts
    (g) Flow slides (involving collapse of loose
        structure)
    (h) Catastrophic debris flows
    (i) Mudflows
        (i) climatic
        (ii) volcanic (lahars)

    Classes (e) to (i) may be referred to as debris
    flow by some classifications.

4. Toppling failures

5. Falls

6. Sub-aqueous slides
    (a) Under-consolidated clay slides (slumps)
    (b) Flow slides

D. FREEZE-THAW PHENOMENA

1. Stone stripes (a form of patterned ground)

2. Periglacial solifluction
    (a) Sheets
    (b) Lobes
    (c) Stone streams

3. Cambering and valley bulging

Table 2.8   Abbreviated Classification of Slope Movements (Varnes 1978)

TYPE OF MOVEMENT		BEDROCK	ENGINEERING SOILS	
			Predominantly coarse	Predominantly fine
FALLS		Rock fall	Debris fall	Earth fall
TOPPLES		Rock topple	Debris topple	Earth topple
SLIDES	ROTATIONAL — FEW UNITS	Rock slump	Debris slump	Earth slump
	TRANSLATIONAL — MANY UNITS	Rock block slide	Debris block slide	Earth block slide
		Rock slide	Debris slide	Earth slide
LATERAL SPREADS		Rock spread	Debris spread	Earth spread
FLOWS		Rock flow (deep creep)	Debris flow (soil creep)	Earth flow
COMPLEX		Combination of two or more principal types of movement		

tact with surrounding material, there is evidence that flow failure does occur sufficiently often for it to warrant inclusion as a separate group.

There is a question then as to the appropriate designation for slope movements which are initiated by shear failure on distinct boundary shear surfaces but which subsequently achieve most of their translational movement by flowage. Some slope movements belonging to this contentious category are illustrated in **Figures 2.5, 2.6, 2.7 and 2.8.** The answer to this dilemma depends on whether the principal interest rests with analysing the conditions of failure or with treating the results of movement. Hutchinson's classification appears to be related more closely to this first purpose.

Hutchinson's classification establishes secondary groups by employing other movement criteria including depth, direction and sequence of movement with respect to the initial failure. These in turn are subdivided in most cases by geotechnically significant features such as material, number of units, and form and depth of displacement. Detailed subdivision within Varnes' classification on the other hand is achieved systematically by adhering rigidly to a material parameter, providing subgroups (bedrock, debris and earth) based on the nature of source material (for example, compare **Figures 2.9 and 2.10**). This approach has been standardised since the 1958 version of the classification by strictly defining the diagnostic characteristic of the material groups and by restricting the parameter to description of the source material rather than the displaced material (which may derive its characteristics as a result of movement). The specific naming of particular classes of slope movement within each subgroup depends on a number of characteristics including degree of saturation and particle size.

Both Hutchinson's and Varnes' classifications have tended to converge over recent years, particularly in terminology. However, some differences still persist and these are pointed out in **Table 2.9.** Whereas Varnes' scheme is perhaps easier to apply and requires less expertise to use, Hutchinson's classification has particular appeal to the engineer contemplating stability analysis. Whatever system is chosen, it is important for the user not only to acknowledge the source of his terminology but also to realise that the dividing lines in most classifications represent an indistinct transition in nature.

Figure 2.5   Fluid 'earth flow' (Varnes' nomenclature)
showing a small surface of rupture, flow track
and accumulation zone, Wairarapa, New Zealand

Figure 2.6   Accumulation zone of a fluid 'earth flow', Wairarapa, New Zealand

Figure 2.7   Tensional zone at the head of a deep, viscous 'earth flow', Otago Peninsula, New Zealand

Figure 2.8   Compressional zone at the toe of a shallow, viscous 'earth flow', Hawkes Bay, New Zealand

Figure 2.9   The Hope 'rock slide' (British Columbia)
              triggered by earthquakes in 1965 involved over
              100 million tonnes of rock, and overwhelmed
              three vehicles, killing the occupants

Figure 2.10   Shallow 'debris slides' involving soil and weathered mudstone, triggered by intense rainstorms, Weber County, New Zealand

Table 2.9   **Correlation of Classifications for Landslides in Engineering Soils**

Hutchinson (1977)	Varnes (1978)	Campbell (1951)
fall	debris fall earth fall	
toppling failure	debris topple earth topple	

single rotational
  slip
successive rotational
  slip
multiple rotational
  slip
	debris slump
	
	earth slump
		slump

spreading failure	earth lateral spread	
graben slide progressive non-circular slide	earth block slide	
debris slide	debris slide debris block slide	earth slip
mudslide bog slide	earth slide	soil slip
catastrophic debris flow	debris flow	debris avalanche
mudflow flow slide bog flow bog burst	earthflow	earthflow mudflow

CAUSES OF INSTABILITY

3

3.1   THE CONCEPT

Slope instability is the condition which gives rise to slope movements. In every slope there are forces (more accurately described as stresses) which tend to promote movement (shear stress) and opposing forces which tend to resist movement (resistance or shear strength). 'Stable' slopes have a margin of stability equal to the excess of resistance over shear stress. Slopes at the point of movement, on the other hand, have no margin of stability, and resistance and shear stress are approximately equal. 'Instability' in any slope represents the condition in which its margin of stability can be reduced to zero. Thus instability is determined by both the inherent margin of stability or the existing slope and the magnitude of transient forces (generally originating outside the slope) which may occur to reduce that margin. This functional definition of instability (as discussed later) may be qualified in terms of the magnitude of instability under consideration, the time period to which it refers, and the degree of certainty with which it can be established. In this chapter 'instability' is used with respect to 'discrete slope movements' (synonymous with 'landslides') as opposed to other forms of mass movement such as creep and subsidence.

Although there is a clear distinction in terms of activity between slopes which are undergoing movement and those which are static, it oversimplifies the issue to classify slopes as being either 'stable' or 'unstable'. If a classification is required, it is better to view slopes as existing in one of three states: stable, unstable (marginally stable), and actively unstable. These three states represent successively smaller margins of stability, culminating in the actively unstable slope where the margin is zero and movement occurs.

In deterministic terms, these three states can be defined (at least theoretically) by the ability of transient forces to produce failure. Stable slopes are those where the margin of stability is sufficiently high to withstand all transient forces. Marginally stable slopes are those which will fail at some time in response to transient forces attaining a certain level of activity. Finally, actively unstable slopes are those in which transient forces produce continuous or intermittent movement.

Unfortunately the distinction between stable and marginally stable states in these terms is more easily made in theory than it is in practice. Besides the problem of establishing accurately the stress conditions within the slope, the main difficulty lies with determining the full range of stress changes that can be brought about by transient factors.[1] Transient factors, such as climate and earthquakes, vary greatly with time and thus a long record is required in order to predict confidently their full range of activity. Without such a record, the rigid deterministic definitions used above cannot be applied and it may be necessary to resort to more approximate probabilistic statements on slope stability. By definition, marginally stable slopes can be predicted to fail at some time within a given period under the prevailing regime of transient forces. The frequency with which critical conditions are reached obviously becomes an important consideration, as even probabilities of failure cannot be stated without such information. Thus frequency assessment also requires a lengthy period of observation.

Identification of the three stability states may also be approached through the history of slope movements. If there is no evidence of past or present movement (for example, cracks, scars or landslide deposits) the slope can be considered to have withstood the activity of transient factors for a long period - at least long enough for any evidence of previous failure to have been obliterated by other slope processes. In the absence of stress analysis (often the case in large-scale regional surveys), such evidence is generally accepted as indicating a stable slope. The problem of this kind of evidence is its inability to reveal deterioration in the margin of stability over time. Certain slopes become weaker with time and their margin of stability may drop sufficiently to allow a 'first time' failure to occur. Slope deterioration, of course, also affects the long-term validity of stability assessment by stress analysis. In marginally stable slopes, however, the history of slope movement may

---

[1]It is necessary here to explain the use of the term <u>factor</u> with respect to slope stability. All forces determining stability are controlled or influenced by identifiable phenomena which are referred to as stability factors. When these operate to induce instability they are referred to as destabilising or causative factors (**Figure 3.1**).

a. PLACE OF OPERATION

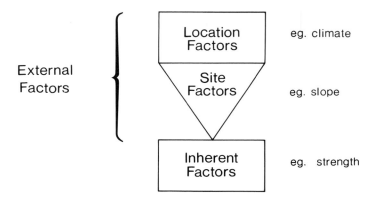

b. FUNCTION

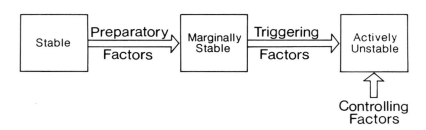

c. RATE OF CHANGE

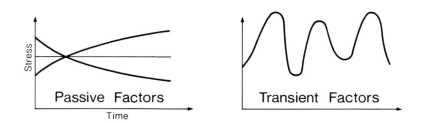

Figure 3.1   Types of causative factors

offer a better approach to determining frequency of failure than does inference by stress analysis because the geological record of critical conditions is often much longer than is the instrumental record. Establishment of the historical record requires the application of dating techniques and sometimes drill-core sampling and therefore can be as costly to implement as stress analysis.

In addition to the geological deterioration of the margin of stability with time, any long-term survey of slope stability must also take into account the possibility of regime changes within the transient factors. Climatic change, shifts of tectonic activity, and human modification of slopes, for example, may all affect both the inherent margin of stability as well as the frequency and magnitude of transient activity. Designations of instability therefore require qualification by time and space.

An excellent example of stability assessment by way of historical record is provided by the work of Hutchinson and Gostelow (1976) in their study of an abandoned cliff in London Clay at Hadleigh, Essex. From the documentary record of damage to Hadleigh Castle and surrounding structures as well as from the stratigraphic relationships of dated landslide deposits, they conclude that the slope had undergone four discrete phases of active instability since its abandonment about 10 000-15 000 years ago. The importance of regime changes in transient factors is also demonstrated, as each phase of landslide activity can be at least partly explained in these terms. The first phase, ending about 10 000 BP occurred under periglacial conditions; the second phase between 7000-6500 BP and the third phase between 2100-2000 BP have been correlated with relatively wet conditions prevailing at the beginning of the Atlantic and Sub-Atlantic periods respectively. The fourth phase, beginning about 1890 AD, although associated with a general climatic cooling, is thought to be due in part to human interference. These regime changes have been set against a background of both slow progressive failure and weathering-related weakening of slope material. The history of this area nicely demonstrates that the stability state of a slope can change through time in response to both internal and external factors.

The concept of three stability states offers a useful framework for understanding the causes of instability. In this context three groups of destabilising factors can be identified on the basis of 'function' (**Figure 3.1b**):

1. preparatory factors which dispose the slope to movement; that is, factors which make the slope susceptible to movement without actually initiating it and thereby tending to place the slope in a marginally stable state;

2. triggering factors which initiate movement; that is, those factors which shift the slope from a marginally

stable to an actively unstable state;

3. controlling (perpetuating) factors which dictate the condition of movement as it takes place; that is, factors which control the form, rate and duration of movement.

A particular factor may take on any of these functions, depending on its degree of activity and the margin of stability within the slope.

Stability assessment by stress analysis is largely dependent on values controlled by the preparatory factors. However, if it is to be an accurate representation of the field problem, the analysis must also anticipate and incorporate the transient conditions likely to be imposed by triggering factors. Certain triggering factors are sufficiently accidental as to defy prediction but many can be characterised by some form of frequency/magnitude analysis and thereby lend themselves to assessment in probability terms.

The same problems face stability assessment that is based simply on historical evidence and comparison with analogous situations (referred to as assessment by precedent). Whatever the method of assessment, an absolute forecast of stability is at present impossible; the assessment of stability is rather a question of probability. As Taylor et al. (1977) observe: 'There is no such thing as a "100% safe" slope, whether natural or man-made. Probabilities of failure form the only rational basis for assessment'.

Superimposed on this degree of uncertainty is the arbitrariness of the decision on a socially acceptable or tolerable level of probability of landslide occurrence. Although the state of the art prevents definitive predictions of stability, a definitive decision is nevertheless often required in respect of human land use or environmental significance. The probability level chosen to categorise instability in this context depends on the purpose of the exercise and the time scale under consideration.

In the social context, for example, instability may be defined with reference to the degree and frequency of landslide activity which presents problems to human affairs, threatens life or property or in general has the potential for influencing land use decisions. In particular, the probability level which might be used to denote the boundary between stability and marginal stability will be strongly influenced by the consequences of any anticipated slope movement - high potential loss areas will require marginal stability to be defined at lower probability levels. In an ideal planning situation, however, judicious management would allocate the most sensitive activities and developments to land which has the lowest probabilities of failure.

Instability is seen in yet a different light by geomorphologists who are interested in landform evolution. To

them instability represents a geologically short-lived condition in which slopes tend to be reduced in mass, height or
angle as a result of some perturbation of geological or
environmental conditions. Instability is thus recognised and
gauged with respect to its effect on landforms. Large-scale
slope movements, by subduing relief, rapidly destroy the conditions necessary for its operation. Unlike many geomorphic
processes, landslide activity is a self-annihilating process
of landform adjustment which tends to give way in time to conditions where form and process take on a more stable and long-
lasting relationship. Where geomorphologists differ amongst
themselves is on how much emphasis should be placed on the
long-term progression towards this more stable and environmen-
tally characteristic end-point and how much should be placed
on the definable and currently measurable relationships and
interactions. In reference to slope development, Brunsden and
Thornes (1979) argued that 'the time needed for adjustment (to
a characteristic form) appears to be of similar order of
magnitude to that needed for changes to the external
controlling variables, such as climate or base level (for
example post-glacial time), so that the characteristic form
concept seems to be a valid and applicable position to adopt
as a basis for landform change studies, at least for the less
resistant systems.' However, they also identify other propo-
sitions which serve to give instability a different degree of
geomorphic significance. For example, they recognise a transient condition in some geomorphic systems where perturbations
in external controls are the norm rather than the exception.
Slope instability in this case would be seen as a recurring
characteristic of the system rather than the mechanism of
adjustment to a perturbation in the long-term trend.
     In the light of these different perceptions of instabi-
lity it can be expected that satisfactory explanations of the
causes will also be qualified by the scale and purpose of concern.

3.2   PRINCIPLES OF CAUSATION

The search for the cause of an individual landslide or an
attempt to designate the state of instability may be prompted
by a need to find an efficient way to respond to the problem,
by legal necessity, or simply by a desire for knowledge.
There are some difficult, almost philosophical, questions
imposed when it comes to accepting, if not establishing, the
causes of mass movement. These relate not only to the analytical difficulties of identifying factors and determining
their relative magnitude and sequence of operation but also to
the evaluation of the relative significance of function compared to magnitude.
     Some of these issues can be examined by resorting once

more to the concept of three stability states identifiable along a spectrum of increasing probabilities of failure. For a slope to move from a stable or even marginally stable state to an actively unstable state, changes must take place which affect the distribution of resistance and shear stresses. One way of visualising this process is to resort to a set of scales as an analogy, with one side representing shear stress and the other resistance. Resting on the 'resistance' pan are several bricks symbolising equal units of stress which must be overcome to produce movement while on the other pan the bricks represent units of stress tending to promote movement. Each brick equates with the stress controlled by one particular stability factor. When the balance is loaded so that it is heavily weighed down on the resistance side, the situation is stable. When the scales approach the point of balance, it is marginally stable and finally, active instability (slope movement) is represented by the scales being over-tipped in favour of the shear stress side. In the marginally unstable state (near the point of balance), as on the real slope, movement could be produced by either removing a 'resistance brick' or by adding a 'shear brick'. Factors which lead to a reduction of shear resistance are referred to by Terzaghi (1950) as 'internal causes' and those which produce an increase in shear stress are classed as 'external causes'.

Assume for the moment that there are four bricks on each side; the scales are balanced and subsequently movement is initiated by the action of a stability factor adding a fifth brick to the shear stress side. Which of the five bricks on the shear stress side caused the movement - or which of the five factors was responsible for the landslide? It can of course be reasonably stated quantitatively that all the bricks are equally responsible for the action, as it is the sum of their component weights which is important. However, before the fifth brick was placed no movement had actually occurred and therefore in terms of direct action the fifth or triggering brick must assume some particular 'functional' significance - yet the fact remains that without the influence of the original bricks the fifth brick would have no special significance.

The action of a triggering factor therefore only partially explains the cause of a landslide; in fact much more significant, destabilising factors may have brought the slope into a condition in which a minor change of regular occurrence could precipitate movement. Indeed there are many components of the slope system which can change independently to destabilise the slope but the significance of any change is dependent on the aggregated effect of the other components. Thus, although it may be possible to identify a single triggering factor, an explanation of the ultimate cause of a landslide invariably involves a number of factors.

As most slopes are stable or at least marginally stable

for most of the time, an actual landslide represents a transient condition infrequently attained by the slope. In the search for destabilising factors then, attention is focused on those factors within the slope system which display the greatest rates of change. An examination of the temporal variability of factors identifies some as being passive (slow changing) and others as being transient or active (fast changing) (**Figure 3.1c**). For example, decrease in rock strength by weathering or increase in slope height from natural erosion are slow changing factors compared to fluctuation in the amount of water in the slope. In the natural setting, it is usually these transient factors which initiate movement. However, in theory, when a slope is in the stable state, by definition, even the normal variation in transient factors is insufficient to produce movement. On the other hand, the conditions which initiate movement on a marginally stable slope can be summarised as follows:

1. a transient factor assumes an extraordinary value;
2. passive factors have gradually (or occasionally abruptly) assumed a state which would allow the normal fluctuation of transient factors to be sufficient to trigger movement;
3. normally passive factors transiently surpass a critical state; for example, rock strength during an earthquake or artificial undercutting increasing the slope angle.

Although passive factors may progressively change over a long period of time to reduce the resistance/shear stress ratio, almost always a transient factor can be identified as having triggered movement. In New Zealand, for example, Selby (1979) observed that: 'Nearly all examples of regional landsliding in recent years have occurred during individual storms or as a result of prolonged wet periods'.

3.3   ACTION OF CAUSATIVE FACTORS

3.3.1   RESISTANCE AND SHEAR STRESS

As explained earlier, a quantitative comparison can be made of the stresses which tend to disturb (shear) the slope material and those which offer resistance. The influence of these two groups of stresses is conventionally expressed by the ratio of resistance to shear stress. This ratio provides a 'factor of safety' which is assumed to yield a value of 1.0 (resistance equals shear stress) at the moment movement occurs and, on the one slope, higher values will represent progressively more stable situations. This method of assessing stability is referred to as 'limiting equilibrium analysis'.

It is necessary to understand how the terms used in

limiting equilibrium analysis relate to each other before they can be linked to the influence of inherent and external stability factors. Thus it is instructive to express the factor of safety in terms of its specific components:

$$\text{Factor of safety} = \frac{\text{resistance}}{\text{shear stress}} = \frac{s}{T} = \frac{\text{shear strength}}{\text{shear stress}}$$

$$= \frac{c + (\sigma - u)\tan\emptyset'}{T} = \frac{c + (\frac{W}{A}\cos B - u)\tan\emptyset'}{\frac{W}{A}\sin B}$$

$$(3.1)$$

Where s = shear strength
T = shear stress
c = cohesion with respect to effective normal stress
σ = total normal stress
u = porewater pressure
$\emptyset'$ = angle of internal friction (shearing resistance) with respect to effective normal stress
W = weight of material; that is, γ V
A = area of shear plane
γ = bulk density of slope material
V = volume of slope material involved
B = angle of the surface on which movement occurs (surface of rupture or 'shear plane')
σ' = σ - u

Resistance is represented here in its simplest form by the Coulomb-Terzaghi shear strength equation (Terzaghi and Peck 1967) but other terms are sometimes included to give a more accurate expression of resistance. This equation indicates that the type of material involved is important in providing two identifiable strength parameters: cohesion and internal friction. Cohesion (c) is the amount of strength offered against shear stress, independent of normal stress. In other words cohesion represents the inherent strength of a material that exists irrespective of any weight imposed directly on (normal to) the surface along which movement tends to take place. The angle of internal friction ($\emptyset$) on the other hand, is the value which indicates the extent to which 'friction', induced by the weight of material acting directly at right angles to the shear plane as a normal stress, contributes to shear strength. Thus the amount of strength derived from internal friction is a product of the angle of internal friction and the normal stress.

The effect of gravity expressed by the weight of material (W) is a fundamental determinant of stability conditions as it not only provides resistance through internal friction but it

Example	Shear plane angle (B)	Sin	Cos	σ (Pa)	T (Pa)
A	30	0.50	0.87	75	131
B	45	0.71	0.71	106	106
C	60	0.87	0.50	131	75

Volume = 100 m^3 Area of shear plane = 10 m^2
Bulk density = 1.5 kg/m^3 Vertical stress = 15 kg/m^2 (150 Pa)
Total weight = 150 kg   (to convert kg/m^2 to Pa, multiply by 10)

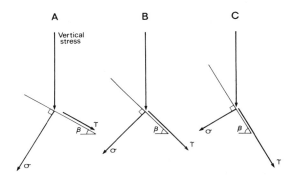

Figure 3.2   Distribution of stress at different angles

also supplies the major disturbing force by producing shear stress. The magnitude of these opposing stresses is determined by resolving the vertically acting weight, expressed as stress (weight divided by the area on which it operates), into two resultants: one acting at right angles to the shear plane (normal stress) and the other acting parallel to the shear plane (shear stress). How much of the weight is manifest as normal stress and how much as shear stress depends on the angle of the shear plane. As shown by the three examples in **Figure 3.2**, normal stress is a function of the cosine of the shear plane angle and shear stress is a function of the sine of that angle. In other words, when the angle increases, the weight provides proportionally less normal stress and correspondingly greater shear stress. The magnitude and direction of stress can be shown diagrammatically with the length of lines being drawn at a scale proportional to units of stress.

The remaining term in the Coulomb-Terzaghi equation is porewater pressure (u). Porewater pressure may be positive, resulting from a build up of groundwater above the shear plane, in which case it acts to reduce the normal stress calculated from total weight. Positive porewater pressure at any point in a freely draining slope is determined by the height of the water table vertically above that point and the unit weight of water. It exerts an upthrust which, by reducing normal stress, detracts from the resistance within the slope. This upthrust will be enhanced if groundwater within the slope is under artesian pressure. Under unsaturated conditions, porewater pressure may have a negative value resulting from tension exerted by attached water and hence may provide an increment of strength.

It must be kept in mind that values obtained for this kind of analysis are determined using samples which represent only a minute fraction of the material involved. Therefore an analysis based solely on such factors, although indicating the extent to which the sample of material is disposed to movement, represents only a point in time and little more than a point in space on the slope in question.

Given the very localised and specific relevance of this kind of limiting equilibrium analysis it is clear that the accuracy and appropriateness of its findings for the whole slope depend on the homogeneity of conditions. In the case of engineered slopes, design criteria and construction standards may ensure an evenly compacted, uniform substrate and consequently laboratory testing conditions can be adjusted to replicate faithfully known properties and thus provide representative values for c and $\mathcal{B}$. Natural slopes, however, present an entirely different set of conditions where variability of properties (Lumb 1966) and difficulty of sampling render the traditional form of analysis highly suspect. In most slopes detailed sampling produces a wide variety of c and β values, permeability varies dramatically and fissures and surfaces may traverse the slope causing anomalous porewater pressures to be recorded. The existence of unforeseen structural discontinuities will often dictate the surface of movement more accurately than will any theoretical determination. It is the resistance generated on such a surface rather than strength properties determined from samples taken on either side of it, which is critical to stability and yet a surface may defy recognition let alone adequate testing procedures.

The problems inherent in applying limiting equilibrium analysis led Hawley (1981) to the view that:

> ...for most natural slopes even the best sampling and testing operation would lead to a scatter of values of c and β which would be more than sufficient to span the range between 'should have failed' and 'stable'. The more thorough the investigation, the greater is the range likely to be....it is not generally practicable to assign meaningful values of c and β to slopes in natural ground.

Although the application of this form of analysis is fraught with difficulties, its theoretical representation of strength and shear stress is sound and can be used to explain how many natural and human factors can cause movement. Movement comes about as a result of change in stability factors and their consequent effect on the distribution of stress. Those factors critical to stability and subject to change are discussed in succeeding sections of this chapter. Before these factors are discussed individually however, brief mention should be made of how their influences have been incorporated into various forms of quantitative stability analysis.

3.3.2 QUANTITATIVE STABILITY ANALYSIS

Earth scientists and engineers are very often faced with specific questions on the stability of a slope. For example: what is the steepest angle at which a slope can be cut without causing a landslide; or what height/slope angle combinations would provide a certain factor of safety for a given slope? To answer these questions in terms of the limiting equilibrium approach outlined, tests would have to be made of strength parameters and hydrological conditions in the slope. In addition, a decision would have to be made, often by trial and error calculations, as to the likely position of the shear surface. On occasions, analysis is carried out after failure has taken place and consequently the position of the shear surface is known but the other factors have to be reconstructed. 'Back analysis' after the event may be carried out to determine which factor varied sufficiently to be responsible for converting the slope to an actively unstable state. Conventional procedures are available and have been used with varying degrees of success to answer these sorts of questions on slope stability.

It is only within the last 100 years or so that quantitative methods of slope stability analysis have been developed and although they have been progressively elaborated, those in general use still employ certain simplifying assumptions that may not always apply. The most common of these is that slope material behaves as a rigid-plastic substance; that is, it can withstand only a certain amount of stress before it deforms irretrievably or ruptures. Another general assumption is that the strength parameters measured either in the laboratory or in the field remain unchanged irrespective of the amount of strain (deformation) that takes place.

Early methods of analysis used the concept of a factor of safety which was treated in deterministic terms but more recently the natural variability of conditions has been accommodated by employing statistical probability theory which provides a measure of the 'likelihood' of failure.

Detailed procedures for slope stability analysis are beyond the scope of this book and interested readers are referred to geotechnical engineering texts such as that by Lee et al. (1983). It should be kept in mind, however, that new information on the behaviour of material and the mechanics of mass movement is continually being incorporated into analytical procedures. The choice of any method of testing or procedure for analysis is best made with a comprehensive understanding of the salient developments in the subject. Some of these are listed in historical order as follows:

1. development of the principal shear strength model (Coulomb 1776 : Mohr 1914);
2. determination of shear plane position, limiting stability and horizontal stress in rock (Rankine 1857);
3. analytical procedure for wedge-shaped failure with plane passing through toe of slope (Culmann 1866);
4. analytical procedure for circular shear failure. 'Method of slices' - able to accommodate inhomogeneity among sample slices; side forces are assumed parallel to shear plane (Fellenius 1927);
5. differentiation of 'total' stress and 'effective' stress and accommodation within the Coulomb-Mohr Law (Terzaghi 1936). This has led to the necessity for a range of testing procedures, determination of porewater pressure and the distinction between 'long-term' and 'short-term' failures (Skempton 1964);
6. friction circle method for rotational landslides in homogeneous soils. Also the investigation of limit state assumptions, log-spiral shear surfaces and the

provision of charts for determining critical shear
surfaces using total stress (Taylor 1948);

7. infinite slope analysis for shallow translational
    slides on planar surfaces (Henkel and Skempton 1954);

8. improvement on Fellenius' analytical procedure for
    circular arc failure (Bishop 1955) - side forces
    treated as acting horizontally;

9. inclusion of effective stresses into circular arc
    method of slices (Bishop and Morgenstern 1960). Sta-
    bility charts provided;

10. methods for taking into account the degree and orien-
    tation of rock joints in slope resistance (Terzaghi
    1962);

11. recognition of 'peak' and 'residual' strength. This
    highlighted the need for site assessments to take
    into account consolidation and erosional history of
    slope material as well as orientation of shear plane
    with respect to orientation of soil fabric (Skempton
    1964);

12. analytical procedures for failure in anisotropic
    soils - that is, soils with distinctively oriented
    structures (Lo 1965);

13. incorporation of interslice forces (Spencer 1967) and
    tension cracks (Spencer 1968) into Bishop's method of
    analysis. Stability charts available;

14. inclusion of composite shear surfaces into stability
    analysis (Janbu 1973; Morgenstern and Price 1965).
    Stability charts provided;

15. application of dynamic stress (relevant to earthquake
    triggering) as opposed to the conventional static
    stress in stability analysis (Seed 1966);

16. departures from the conventional assumption that
    stress/strain behaviour of material is of the rigid-
    plastic type. Including aspects of fluid mechanics
    in analysing flow movement (Johnson 1970; Cunningham
    1971; Hutchinson and Bhandari 1971);

17. inclusion of seepage forces in infinite slope
    analysis for shallow landslides (Hartsog and Martin
    1974);

18. probabilistic approaches in slope stability analysis
    (Lumb 1974; Ward et al. 1981).

3.3.3   VARIATION IN WEIGHT

The first of many stability factors to be discussed individually is variation in weight. Changes in the weight imposed
on a slope (loading) result from both natural and human agencies and may alter the degree of stability depending on pre-
existing slope conditions. In some cases, for example, an
increase in weight (surcharge) may increase stability; in
others it may decrease stability.

Natural processes producing a variation in weight include:

1. precipitation (rain, snow and ice), runoff, ponding, evapotranspiration and drainage;
2. mass movement deposition and erosion;
3. deposition and erosion from other geomorphic processes;
4. extrusion of volcanic material;
5. overthrust faulting;
6. vegetation growth and vegetation destruction;
7. seepage drag from percolating water;
8. variation in atmospheric pressure.

Human agencies responsible for variation in weight include:

1. storage and conveyance of water and other fluids (pipes, tanks, canals, reservoirs);
2. leakage of water and fluids from storage and conveyance methods;
3. land development earthworks including roading (cuts, fills, side-cast);
4. industrial activity including quarrying and mining (stockpiles, waste piles, overburden and resource removal);
5. afforestation and deforestation;
6. machinery (mobile and static);
7. erection of buildings and other structures.

The relative magnitude of a few examples of surcharge is illustrated in **Table 3.1**.

Table 3.1   Examples of Surcharge

Source	Loading (kPa)
10 cm thick concrete slab	2.3
Spruce/Hemlock forest	2.4
Annual range of atmospheric pressure	5.0
1 metre of dry soil	10.0
1 metre of gravel[1]	10.8
1 metre of 30% moist soil[2]	13.0
One storey concrete building	18.5
1 metre of saturated soil	20.0
1 metre thick basalt flow	30.0

Note: 1 assuming a void ratio of 0.6 and specific gravity of solids at 2.7
       2 assuming porosity of 50%

The effect of surcharge on stability depends on the stress/strain properties of the slope material (which are affected by rate of loading, pressure and temperature (Billings 1972)), its permeability and in particular the presence or absence of cohesion.

Cohesionless slope. In the case of a cohesionless, well-drained, competent material, such as completely jointed sandstone, resistance will be represented simply by:

from Equation (3.1) $\qquad s = \sigma \tan\theta$ $\qquad$ **(3.2)**

or $\qquad s = \dfrac{W}{A} \cos B \tan\theta$ $\qquad$ **(3.3)**

and shear stress by $\qquad T = \dfrac{W}{A} \sin B$ $\qquad$ **(3.4)**

In cohesionless material it is thus evident that the relative proportions of shear stress and resistance are determined by the angle of the shear plane and the angle of internal friction and are independent of the weight. If surcharge is imposed and B and θ remain the same then, although the absolute values of s and T increase, their ratio (factor of safety) remains unchanged. For example, the effect of a fourfold increase in vertical stress upon the shear plane produces the following results:

Given $\dfrac{W}{A}$ = 10 kPa, θ = 40° and B = 30°

from Equation (3.3) then $\quad$ s = 10 x 0.87 x 0.84
$\qquad\qquad\qquad\qquad\qquad$ s = 7.31 kPa

and from Equation (3.4) $\quad$ T = 10 x 0.5
$\qquad\qquad\qquad\qquad\qquad$ T = 5 kPa

so factor of safety $\quad (\dfrac{s}{T})$ = 1.46

When $\dfrac{W}{A}$ is increased to 40 kPa, then:
$\qquad\qquad\qquad\qquad\quad$ s = 40 x 0.87 x 0.84
$\qquad\qquad\qquad\qquad\quad$ s = 29.23 kPa

$\qquad\qquad\qquad$ and $\quad$ T = 40 x 0.5
$\qquad\qquad\qquad\qquad\qquad$ T = 20 kPa

so factor of safety $\quad (\dfrac{s}{T})$ = 1.46

This shows that increase in weight has no effect on the factor of safety in cohesionless material; that is, the change in weight is unable to produce movement. However, the four-fold increase in vertical stress has meant, in absolute terms, that the magnitude of stress required to cause movement has increased with an increase in surcharge. This magnitude is measured by the 'excess strength' (s - **T**) (referred to earlier as the margin of stability) which, in this example, was originally 2.31 kPa and which increased to 9.23 kPa as the vertical stress increased four times.

The 'excess strength' on an initially stable slope can be represented as:

$$(\frac{W}{A}) = [\cos B . \tan\emptyset - \sin B] \qquad (3.5)$$

or

$$(\frac{W}{A}) = \cos B [\tan\emptyset - \tan B] \qquad (3.6)$$

An additional weight (ΔW) on the shear plane would thus produce an increment in the 'excess strength' equal to:

$$\Delta (s - \mathbf{T}) = (\frac{\Delta W}{A}) \cos B (\tan\emptyset - \tan B) \qquad (3.7)$$

Because $\emptyset$ > B, which is the only possibility on a stable cohesionless slope, each increase in weight will produce a positive increment in 'excess strength' irrespective of the slope angle.

However, in certain circumstances, the mechanical effect of tree roots may provide sufficient strength to retain an otherwise cohesionless slope at least temporarily at an angle greater than the angle of internal friction. In such a case surcharge will reduce the margin of excess strength.

In summary, changes in weight on a cohesionless slope cannot in themselves cause a landslide but they do affect the susceptibility of a slope to triggering by some other factor. Increase in surcharge provides an increased margin to strength which would therefore require a larger disturbing force to trigger movement. The example used to illustrate the effect of surcharge also shows that careful interpretation of factors of safety is required when comparing a number of different slopes. This is because they indicate 'excess strength' only as a proportion of the existing shear stress and, in themselves, give no indication of the absolute magnitude of that 'excess strength'. It is the magnitude of the excess strength which should be measured against the magnitude and frequency of the likely disturbing forces in order to provide a realistic indication of the safety of the slope.

Cohesive slope.   In the case of a slope comprised of cohesive material, the influence of surcharge is quite different. Increase in weight on a cohesive slope with a pre-determined potential shear plane will reduce the factor of safety and, if of sufficient magnitude, will produce failure. At the same time there will be either a decrease or an increase in the 'excess strength' depending on whether $\mathscr{B}$ is less than or greater than B. For example, the effect of a four-fold increase in vertical stress, when $\mathscr{B} < B$, is as follows:

Given $\dfrac{W}{A}$ = 10 kPa, $\mathscr{B}$ = 20°, B = 45° and C = 20 kPa

then from Equation (3.1)   s = 20 + (10 x 0.7 x 0.36)
                                                    s = 22.52 kPa

and from Equation (3.4)   T = 10 x 0.7
                                                  T = 7.0 kPa

so factor of safety   ($\dfrac{s}{T}$) = 3.22

excess strength (s - T)   = 15.52 kPa

When $\dfrac{W}{A}$ is increased to 40 kPa, then:

                                        s = 20 + (40 x 0.7 x 0.36)
                                        s = 30.08 kPa

                    and   T = 40 x 0.7
                              T = 28 kPa

so factor of safety   ($\dfrac{s}{T}$) = 1.07

excess strength (s - T)   = 2.08 kPa

However, if the above calculations were made with $\mathscr{B}$ = 40° and B = 30° (that is, with $\mathscr{B} > B$) then the increase in weight would reduce the factor of safety from 5.4 to 2.4 but the excess strength would <u>increase</u> from 22.3 kPa to 29.1 kPa.

Other effects. Surcharge may have effects other than a direct mechanical influence on the distribution of stress as described above. Soils, particularly those consisting largely of silt, clay or organic matter, are readily compressed under loading. If these are able to drain under loading, consolidation may occur and significantly reduce void space among the particles (see Means and Parcher 1963 for a comprehensive treatment). Reduction in void space has two opposing effects with respect to the action of stability factors. One of these is the increase in strength attributable to internal friction, as there is generally an inverse relationship between void ratio and the angle of internal friction (Kirkpatrick 1965). The other effect is the increased likelihood of positive porewater pressure being developed as a result of a reduction in pore space, leading to a reduction in strength. The extent to which these changes can affect stability will depend largely on the drainage conditions within the slope.

If groundwater is already within the zone subject to surcharge, sudden loading will prevent drainage and excess porewater pressures will develop, immediately reducing resistance. Hutchinson and Bhandari (1971) have described such 'undrained loading' resulting from debris accumulating on the head of a mudslide as an important mechanism in promoting downslope movement. Their work was prompted by the observation that certain coastal mudslides advanced on slopes flatter than those corresponding to limiting equilibrium for conditions of residual strength and groundwater flowing parallel to the slope surface. From both theoretical considerations and porewater pressure measurements in the field, they determined that debris accumulating at the rear of a mudslide loaded the landslide sufficiently quickly to prevent drainage and to increase porewater pressure to artesian levels. Such pressures appear to induce a forward thrust which may initiate shearing movements or accelerate the rate of movement downslope.

Some soils with high water content, particularly after disturbance, behave as highly viscous fluids and even some normally competent clay and silt-rich sedimentary rocks will exhibit plastic flow under prolonged loading. With such material, surcharge tends to promote downslope flow. Benson (1940) has described a number of localities in eastern Otago, New Zealand where volcanic rock extruded over mudstones during the Miocene has initiated plastic flow of the less competent underlying material (Figure 3.3). The deformation of the mudstone as it is squeezed from under the lava sheets and the sliding of the volcanic material itself has, over a number of years, been sufficient to dislocate buildings, pipelines, roads and railway lines.

These examples show that variation in weight cannot directly instigate failure on cohesionless slopes while surcharge tends to make such slopes less susceptible to failure. Conversely, on cohesive slopes surcharge may produce failure although the excess strength will increase if the angle of internal friction is greater than the angle of slope, and will decrease if it is less than the angle of slope.

Other less direct effects of surcharge relate to the degree of resulting consolidation, its consequent influence on the angle of internal friction and the enhancement of positive porewater pressures. In addition, surcharge on material which has been induced to behave as a viscous fluid and on material with a plastic response can initiate and accelerate movement.

Figure 3.3   The large slump (centre-left) has occurred on the edge of a sheet of Miocene volcanic rock underlain by incompetent mudstone

In certain circumstances, such as those represented by deep-seated earthflows, the acceleration or instigation of mass movement has been attributed to surcharges as small as those resulting from afforestation. The prediction of the influence of surcharge however, depends not only on the degree of loading but on material properties including stress-strain behaviour, permeability, void ratio, cohesiveness, as well as the stratigraphic relationship of these properties.

3.3.4 INCREASE IN SLOPE HEIGHT

Increase in slope height occurs most commonly as a result of excavation and land development or in response to fluvial downcutting. Even if none of the other slope parameters is changed, increased height in some circumstances may be sufficient to initiate slope movement.

As increased height may result in increased weight over a potential shear plane, its influence is dictated by the same factors which determine the influence of weight on stability. Thus with a constant shear plane angle, increase in height may cause failure in a cohesive slope but not on a cohesionless slope.

In discussing hard rock slopes, and drawing largely on the work of Terzaghi (1962), Carson and Kirkby (1972) provide a useful analysis of the factors controlling the critical height (H'_c) above which failure will occur. They argue that the processes which create slope relief invariably involve the removal of material from the slope face and thus a reduction in lateral support for the slope. This reduction in support sets up a zone of negative earth pressure or tension extending for some distance into the slope. At the limit of the zone of tension a near vertical crack may appear, depending on the tensile strength of the material. Terzaghi (1943) has shown that on a vertical slope, and assuming Rankine's active state of stress (essentially assuming failure by material movement away from the face (Capper and Cassie 1969)), the depth of this crack (Z_O) is equal to:

$$Z_O = \frac{2c}{\gamma} \tan (45° + \frac{\theta}{2}) \qquad (3.8)$$

where the terms are defined as for Equation (3.1)

Rankine's assumptions also dictate that with a vertical slope any failure plane would pass through the toe of the slope at an angle equal to 45° from the horizontal plus half the angle of internal friction.

In many rock slides, rock avalanches and slab failures, the failure plane is seen to intersect a tension crack at the back of the landslide. Erosional scars formed in this manner are usually easily identified, as the failure plane exhibits fresh ruptures and slickensides while the inside wall of the former tension crack forms a distinctive crown at the head of the landslide. The surface of this feature is often more weathered than the failure plane, suggesting that tension cracks may anticipate a failure by a considerable period.

The presence of the tension crack provides a physical limit to the area of the potential shear plane as slope height is increased. As the cohesive strength of the slope is equal to c x area of the shear plane, total cohesive strength remains constant during height increase whereas shearing stress continues to rise as a function of the increase in weight above the shear plane. Thus the presence of tension cracks or surfaces of relative weakness within the slope will dictate a maximum stable height that is considerably lower than that which might be achieved in an ideally uniform slope material. The effect of tension cracks (both dry and water-filled) on reducing slope resistance is illustrated in Section 3.3.9 CHANGES IN WATER CONTENT.

However, even in a simplified ideal situation, without the presence of tension cracks, as illustrated in **Figure 3.4,** it can be seen that an increase in height will lower the factor of safety because shear stress increases at a greater rate than does shear strength. In this example height is the only independent variable changed. Within a vertical slope and assuming homogeneity of material, the angle of the potential shear plane will remain constant. It is evident from this example that increase in volume is equal to the increase in height squared whereas increase in the area of the shear plane is only equivalent to the increase in height. This means that when the height of a cliff is doubled, the weight of the material above the shear plane (volume x bulk density) and thus the total stress, increases four-fold. At the same time the area of the shear plane itself is only doubled. As the total cohesion that can be mobilised on the shear plane is equal to c x area of shear plane, a doubling of height represents only a doubling of total cohesion as against a four-fold increase of stress. This difference can also be seen by comparing c and σ on a per unit area basis. In this example c remains constant at 100 kg/m^2 as height increases whereas σ increases at the same rate as height, from 0.71 kg/m^2 to 2.14 kg/m^2 with a trebling of height.

The critical height H'_c that can be achieved for a particular material can thus be determined using the theoretical principles outlined. For a vertical, free-standing column of rock the relationship is summarised by:

$$H'_c = \frac{qu}{\gamma} \qquad (3.9)$$

where qu is a measurement of rock strength referred to as unconfined compressive strength

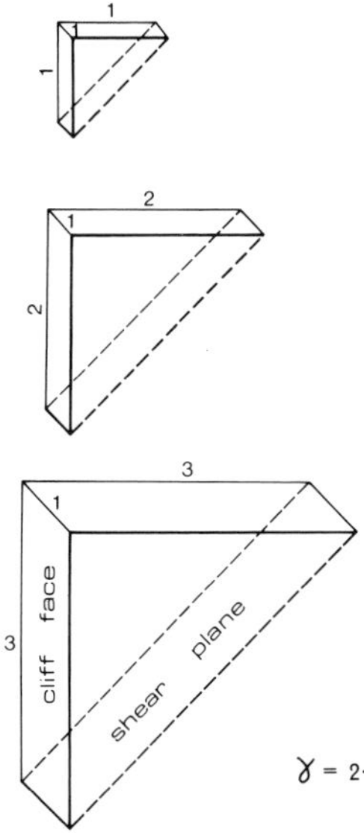
1
1
1
2
1
2
3
1
3
cliff face
shear plane

height m	c kg/m²	σ kg/m²	vol m³	A Shear P m²	weight kg	c × A Total C (kg/m²)	
1	100	0·71	0·5	1·4	1	140	
2	**0**	**2**	**4**	**2**	**4**	**2**	x increase
2	100	1·43	2·0	2·8	4	280	
3	**0**	**3**	**9**	**3**	**9**	**3**	x increase
3	100	2·14	4·5	4·2	9	420	

$\gamma = 2{\cdot}0\ kg/m^3$

Figure 3.4 **Change in value of stress and strength parameters with height**

Unconfined compressive strength is generally obtained by using a simple testing device to exert a principal stress on the end of a laterally unconfined cylinder of rock until it fails. The stress required to achieve rupture is taken as the unconfined compressive strength. Although the formula is theoretically sound, in itself it is much too crude a representation of slope strength to be a useful way of predicting the critical height of natural slopes. As natural slopes rarely achieve a height anywhere near the theoretical maximum, it is obvious that structural discontinuities and variation in rock properties supercede the assumptions of the formula in their influence on stability.

Sampling limitations which reduce the applicability of many laboratory-based predictions have lead to more empirically-based field solutions. **Figure 3.5** shows how field measurements of actively unstable and stable slopes can be used to obtain a measure of critical height. Relationships between height and slope angle are established by measuring a large number of slopes for one particular regional terrain; that is, on the same rock type and in the same climatic regime. Slopes which show signs of instability are distinguished from stable slopes - the height and slope angle parameters for both stable and unstable slopes are plotted on the graph. If these two parameters are critical determinants of stability it will be possible to construct a definitive envelope line separating stable and unstable conditions and showing an inverse relationship between height and angle. A similar line may be constructed by establishing the difference between pre-failure and post-failure slope geometry. As indicated by the position of the envelopes in Figure 3.5, it is often necessary to distinguish further stability conditions resulting from the influence of other factors such as type of mass movement or tectonic history. At a given slope angle, surface slips in boulder clay occur on slopes much lower in height than those on which deep slips are found and mass movement occurs much more readily in shattered than in unfaulted greywacke.

The advantage of this method for determining critical height is that the measurements reflect the sum-total of all the stability influences, both past and present. Rock strength, for example, is not determined in isolation and then integrated with other factors which have also been individually determined. Rather the in situ influence of rock strength over a historical range of climatic and other conditions is measured from an actual landform response. The stability limits determined, however, are conservative as they will represent a response to the 'worst possible case' in the observable history of the slope.

The main disadvantage of this method is that the relationship can be established only in areas which already contain unstable slopes. Without a precise quantitative analysis

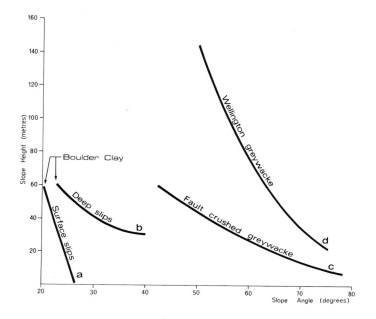

Figure 3.5   **Relationship between maximum stable angle and
height of slope** (after Crozier 1984):
    (a)   stability with respect to surface slips
          and slumps in boulder clay, determined by
          field observations (Skempton 1953);
    (b)   stability with respect to deep slips in
          boulder clay determined by theoretical
          calculation (Skempton 1953);
    (c)   stability of intensely sheared rock
          situated within a zone extending 800 m
          west of the Wellington Fault, New Zealand,
          field observations (Grant-Taylor 1964);
    (d)   stability of the strongest greywacke
          slopes in the Wellington area, New
          Zealand, field observations (Grant-Taylor
          1964).

of individual stability factors, extrapolation of envelope lines between different terrain is not generally justified.

3.3.5 LATERAL SUPPORT AND SLOPE ANGLE

Material which is in contact with the slope or is part of the slope and offers more resistance than shear stress, constitutes lateral support. Generally lateral support is generated by slope material at the toe or base of the slope or in some instances by water and ice bodies or artificial support. Consequently removal of lateral support in the natural setting occurs most commonly as a result of undercutting by coastal, fluvial and glacial processes (Figure 3.6). In urban areas, removal of lateral support frequently occurs as a result of excavation for building sites and access ways. Next to the action of climatic factors, removal of lateral support is probably the most common triggering cause of slope movements. On a geological time-scale, however, removal of lateral support by natural processes and the consequent oversteepening is the principal mechanism in inducing the slope to change from a stable to marginally stable state.

Removal of slope material providing lateral support invariably involves steepening of part of the slope form, change of weight on the shear plane and in some cases a reduction of shear plane area and steepening of the potential shear plane angle, compared to the unaltered slope conditions. As is evident in Figure 3.7 most of these changes will reduce the stability of the slope to the extent discussed in the preceding sections on changes in weight and height. In Figure 3.7A, the relationship is shown between the direction of major principal stress (σ_1) and the orientation of the shear plane (after Carson and Kirkby 1972). Given slope material of uniform strength, the shear plane will develop at a set angle $[45° + (\emptyset/2)]$ to the direction of σ_1 which varies throughout the slope in response to slope geometry. After removal of lateral support has steepened the slope face (Figure 3.7B), at any given distance into the slope, σ_1 is closer to the vertical than it was prior to alteration of the slope. As a consequence, the potential shear plane is relatively steeper over its length indicating that gravitationally-induced strength has decreased while shear stress has increased. Even on a curved shear surface, which is fixed in position by structural factors, the material removed (shaded triangle in Figure 3.7B), by virtue of the gentle angle of its shear plane, represents that part of the slope where more of the unit weight acts to induce friction than it does anywhere else on the potential slide. On some rotational slides the curvature of the shear surface may be sufficient for the shear surface at the base of the slope to dip 'backwards' into the slope. During movement, slide material rides up at the base of the slope, continuously increasing the moments of force (restoring

Figure 3.6 Coastal sand dune deposits subject to recent slumping as a result of removal of lateral support by storm wave action

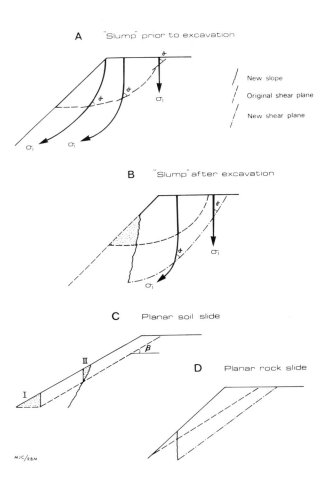

Figure 3.7 **The effect on the position of the shear plane of removing lateral support**

moments) working into the slope. When sufficient material has ridden upwards far enough for the restoring moments to balance the disturbing moments, the material will come to rest. Slides of this sort are very sensitive to removal of lateral support and removal of material from the toe can readily reactivate movement.

In the case of a potential planar slide of soil mantle over resistant bedrock **(Figure 3.7C)**, the new slope formed by excavation will have lost not only the relatively high frictional resistance at the former toe (triangle I) but also an increment weight (triangle II). The frictional resistance generated by triangle I is not matched by any part of the mantle on the new slope and the loss in weight (triangle II) will either reduce or increase resistance depending on whether B is less than or greater than the angle of internal friction.

In **Figure 3.7D**, a planar rock slide is represented with a potential failure plane passing through the toe of the slope. The potential shear plane for the altered slope, assuming a toe failure, will steepen and shorten. As a result, shear stresses will increase while total cohesion will be reduced with a reduction in shear plane area.

Unloading associated with removal of lateral support and the consequent development of landslides has been studied extensively in the coastal setting by Hutchinson (1975; 1976; 1979). He notes that unloading of clay slopes produces expansion and softening of the exposed material, particularly if it is overconsolidated and dilatant. In jointed and fissured material, an expansion of discontinuities produces non-uniform shear strains and enhances the opportunity for the action of joint processes (see Section 3.3.7 JOINT PROCESSES).

The ratio between unloading-induced 'weathering' of clay slopes and the rate of removal of material from the toe of the slope is considered by Hutchinson (1975) to determine the particular type of slope movement which takes place. Under moderate toe erosion, shallow mudslides are present and the erosion is confined almost entirely to the mudslide tongues. Strong toe erosion, on the other hand, allows in situ clay to be affected by removal of lateral support and as a consequence deep-seated rotational landslides develop. This condition is illustrated by the landslide in the Walton cliffs, Essex, which was triggered by a tidal surge on 29 November 1897 and by the Beltinge Cliff 'Miramar' landslide of 4 February 1953. In a situation where there is negligible toe erosion, only shallow rotational landslides occur.

Although human action and erosion are the most common agents for removing lateral support and steepening slopes, over the longer period of geological time, endogenic processes such as tectonic and volcanic activity are ultimately responsible for creating relief. In most landslide problems the landform and structure which result from endogenic activity can be treated as passive factors. However, in certain cases endogenic processes may be sufficiently active to destabilise a slope. For example, freshly oversteepened fault surfaces were exposed in 1855 along many kilometres of the Wairarapa Fault in New Zealand when 12.2 m of horizontal movement and 2.7 m of vertical displacement were recorded. The Rimutaka mountain range lying immediately to the west of this fault, has, in recent geological time, risen to a height of nearly 1000 m largely by a series of such movements. Average uplift by faulting over the last 20 000 years is estimated at 0.5 mm/y (Lensen and Vella 1971) and over the last 6500 years there is evidence for at least five episodic movements having produced a total uplift of 30 m (4 mm/y) (Wellman 1969). The regional tilting associated with uplift along one margin of the Rimutaka block extends at least 15 km westward and has the effect of steepening all slopes with a westerly aspect (that is, facing west). The instability evident on the western flanks of the Rimutaka Range can be primarily attributed to tectonic oversteepening and secondarily to forest degradation as a result of browsing mammals. Triggering events in this unstable terrain have been identified as earthquakes and intense rainstorms (Cunningham and Arnott 1964).

Severe landsliding may also be closely associated with volcanic activity - a recent and probably the best studied example was related to the eruption of Mount St Helens on 18 May 1980 in the Cascade Mountains, Washington State of USA. Magmatic intrusion into the volcano during the preceding month caused bulging and steepening on its northern slopes which had also been weakened by earthquake and hydrothermal activity (Schuster 1982). The resulting collapse of the upper northern slopes is thought to have triggered the major eruption and to have produced a massive (2.8 km^3) hot debris avalanche. The avalanche travelled 22 km down the North Fork Toutle River valley giving way to a mudflow which continued for about 100 km. The damage caused by these flows was immense including the destruction of over 200 buildings, approximately 50 bridges, large extents of roading, railway and other services as well having severe ecological consequences.

3.3.6 REMOVAL OF UNDERLYING SUPPORT

The effects of removal or reduction in the competence of underlying material on the ability of a slope to withstand superincumbent weight of overlying material is easily visualised. As these effects commonly destabilise slopes it is important to be able to identify the location, rate and mechanism of the processes involved.

The processes responsible for the removal of underlying support include:

1. surface action by localised weathering and erosive agents;
2. preferential surface erosion or weathering as a

result of differences in erodibility;
3. localised or diffuse subterranean removal of material by mechanical eluviation, solution or plastic flow;
4. mining, quarrying or withdrawal of fluids.

Surface action. The best known examples of localised surface action of erosional agents is the undercutting caused by the hydraulic action, abrasion and solution of rivers, particularly on outside river bank bends. Wave-cut notches at the base of coastal cliffs occur as a result of similar concentrated processes enhanced by air compression and expansion caused by breaking waves. Preferential weathering resulting from wetting and drying cycles, salt weathering and biological activity tends to concentrate in these localities. Certain soluble rocks are also believed to undergo preferential weathering at the base of slopes or just below the soil surface where active water tends to accumulate and pH values rise. The intensity of preferential erosive and weathering agents can often be correlated with slope aspect. In areas where the availability of water is the limiting factor, maximum activity occurs on shady slopes and where thermal, wetting/drying or freeze/thaw cycles control the rate of activity, the sunny faces usually display the greatest rates. The prevailing or dominant wind direction may also influence the rate of some of these processes, particularly in the case of aeolian abrasion.

There are many well known examples of alternating sequences of rocks of different degrees of erodibility having become susceptible to landsliding. Landforms capped by resistant beds, generally sandstone and limestone and underlain by softer mudstones or siltstones, often have their lower slopes littered with fallen blocks of caprock, attesting to preferential erosion of the underlying material. Quarrying and opencast mining can have a similar but much more rapid effect.

Subterranean action. A more insidious consequence of alternating rock types is their effect in promoting different rates of subterranean removal of supporting material. The worst possible case in this respect is the presence of a bed of water-bearing, cohesionless sand overlying a less permeable stratum. Hutchinson (1982) notes that seepage erosion which actively removes the sand is often initiated by cutting of a free face within such a stratigraphic sequence. The depletion of the sandy layer may continue, given a sufficient groundwater supply, until the superincumbent material collapses. Such sequences are often encountered in volcanic tephra and in other terrestrial deposits, particularly those of glaciofluvial origin. The highest rates of depletion were observed by Hutchinson (1982) to occur in beds containing cohesionless sands which have a median particle size of between 0.09 mm and 0.25 mm.

Localised subterranean erosion within soils is evidenced by the formation of tunnels (also referred to as pipes or under-runners). In embankments, earth dams, and neighbouring natural ground where an hydraulic head of water is created nearby, such features have often been the precursor to rapid deterioration of support and eventual collapse of the structure. In these situations seepage, induced drainage and landform alteration as a result of earthworks produce changes to the water table and soil moisture regime. Certain soils react to this hydrological stress by radically changing volume or losing sufficient strength to induce cracking. Seepage water migrates to these cracks or to the interface between artificial structures and natural ground and localised flow along such pathways eluviates the fine particles thus further reducing underlying support.

This process is illustrated by major failures that occurred in the canal systems of two New Zealand hydro electric power schemes. The first took place in September 1981 when a deep-seated collapse is believed to have occurred in the original ground underlying supporting fill at the power house end of the Ruahihi canal. As a result, one million cubic metres of liquid mud and rubble travelled 1200 m down a tributary valley into a nearby river (Soil and Water 1982). In this and the subsequent Wheao canal collapse there is both direct and indirect evidence suggesting that localised subsurface erosion was one of the important mechanisms contributing to failure (Soil and Water 1983). Extensive cracking and the appearance of a 'sink-hole' had been noted in supporting fill prior to the Ruahihi collapse and in both cases a volcanic ash (which in certain localities contains the clay allophane) had either been used in construction or was present along with other tephra beds in the local substrate. Allophane, which is a weathering product of certain volcanic ashes, is highly responsive to any induced changes in the hydrological regime and to tensional stress brought about by small movements of any supporting structure. Allophanic soil has a high natural water content, is extremely sensitive (suffers loss of strength on remoulding), undergoes irreversible changes on drying (standard laboratory drying procedures are therefore inappropriate), exhibits brittle behaviour and possesses a high potential for shrinkage. In its submission to the committee of inquiry into the Wheao canal failure, the New Zealand Geomechanics Society observed that: 'Leakage from any water-retaining structure into complex volcanic materials, often containing highly permeable cohesionless sands, has the potential for initiating erosion and/or piping failures' (New Zealand Geomechanics Society 1983).

The disastrous collapse of Idaho's 90 m high earthfill dam on the Teton River in June 1976 also indicates how small cracks under high water pressure can lead to rapid piping failure (Clark 1983).

Localised subterranean erosion in natural soils is generally manifest in the form of pipe-shaft or tunnel-gully systems (**Figure 3.8**). Pipes are generally less than one metre in diameter, located within two metres of the slope surface and yield an intermittent flow of storm water. They are most commonly encountered under grassland or bare surfaces and occur either singly or as part of a dendritic network. Collapse of overlying ground produces a shaft leading into the tunnel and subsequent coalescence of shafts and accelerated surface erosion converts the system into open gullies.

The extent to which piping can promote shallow land-sliding is an unresolved issue. In an analysis of the mapped distribution of tunnel gullies in New Zealand, Lynn and Eyles (1981) found that in the 435 000 ha affected, this form of erosion was commonly associated with soil slips and earth-flows. It is also evident that pipes are occasionally exposed in the crowns of fresh landslides, suggesting that under wet conditions a concentrated supply of water would be discharging into or through the potential landslide. Water flow through a pipe can become impeded by roof collapse or other forms of blockage allowing sufficient temporary build-up of groundwater to trigger mass movement (Pierson 1983). However, pipes generally appear to act as an efficient drainage system (Crozier et al. 1982) in unstable terrain.

As Jones (1978) has observed, the removal of material by closely spaced pipes may predispose a natural slope to mass movement by removing underlying support. Maximum reduction in support would be encountered immediately upslope of a confluence within a tunnel network. The extent to which this form of subterranean erosion affects the stability of natural slopes therefore depends on the balance between the negative effects of reduced support and the positive effects of enhanced drainage.

There are five necessary conditions and a wide range of enhancing factors for the development of tunnels (Kerrison 1981):

1. a source of water. This may be provided by streams or impounded water bodies but rainfall is the primary source. Tunnel-gully systems have been found under a wide range of climatic regimes although optimum conditions appear to occur where the annual rainfall is between 500 and 760 mm and where both drought and high intensity rainstorms are experienced;

2. surface infiltration rate must exceed the permeability rate of some subsurface layer. The subsurface layer must have a coefficient of permeability low enough to allow periodic saturation of the overlying soil. Entry of water into the soil in tunnelled areas is often by way of surface cracks promoted by the presence of swelling clays such as montmorillonite or by reduction in insulation brought

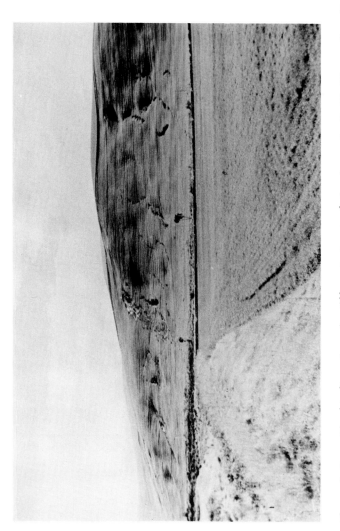

Figure 3.8   Pipe/shaft systems in different stages of development, North Otago, New Zealand

about by a degraded vegetation cover;
3. the presence of an erodible layer above a retarding layer. Erodibility is usually at a maximum in fine sediments (clays, silt, sand, volcanic ash) but tunnel gullies have been found in material as coarse as colluvium. Erodibility is enhanced by the presence of dispersible clays and soils with a high exchangeable sodium percentage or other deflocculant ions. Compact retarding layers may be natural pedogenic pans such as loess fragipans or they may result from certain cultivation practices;
4. the presence of water under a hydraulic gradient. This is a function of water supply and slope. Tunnel gullies have been found on slopes ranging from 10° to 35° but most commonly on slopes between about 15° and 25°;
5. the existence of an outlet for flowing water (**Figure 3.9**). This may be in the form of a tunnel exit, an area of well-developed cracks or highly permeable soil.

Human agencies are also responsible for the subterranean removal of underlying support by underground mining and the removal of gas, oil, and water from within the ground. As these actions are more important in causing subsidence than slope movement they will not be discussed here.

3.3.7   JOINT PROCESSES

The surficial part of a rock mass or soil is affected by external influences and atmospheric conditions which may lead to lateral pressures being experienced. These are pressures which augment shear stress. There are a number of active factors involved operating primarily within joints and other void spaces - their actions are therefore referred to as joint processes. In addition to setting up lateral pressure, these factors are also responsible for both lowering inherent strength and increasing susceptibility to triggering influences. Joint processes are discussed below under the following headings:
1. root wedging;
2. cleft-water pressure;
3. ice wedging;
4. salt wedging;
5. water wedging;
6. clay swelling.

Root wedging. Root wedging occurs primarily on exposed, jointed rock surfaces - if a soil cover is present roots tend to spread laterally rather than penetrate bedrock. Rooting depth is extremely variable ranging from that of spruce trees

Figure 3.9   Small fan of sand at the outlet of a sub-
               terranean pipe

of the taiga (penetrating only 0.3 m) to Ponderosa pine of the Colorado Rockies (achieving depths of 10 to 12 m) (Pitty 1971). Penetration occurs within even the most compact joints and other structural discontinuities, the roots growing in length and diameter with time. Joints are thus forced open to accommodate the increasing biomass thereby destroying the interlocking action amongst the joint blocks and consequently reducing internal friction.

Birot (1962) claims that the force set up by a living root 10 cm in diameter and 1.0 m in length is capable of moving a block weighing about 41 tonnes. This process may be responsible for periodically dislodging blocks on the surface of the rock face (Eyles et al. 1978) but the associated increase in 'apparent cohesion' generated by the root network at least temporarily offsets any reduction in strength resulting from wedging.

On most slopes the presence of a vegetation cover appears to favour stability at least in the short term. However, within the surface zone of steep, intensely-jointed bedrock slopes, root wedging may reduce the angle of internal friction to a value which is actually less than the angle of slope. Consequently, what was inherently a cohesionless slope (because of the presence of intersecting joints) will now be standing precariously, relying on an increment of root cohesion to maintain stability. In this condition even small surcharges such as that resulting from rainfall will act as a decrement to stability and may trigger a shallow landslide. If landslides are occurring under these conditions, paradoxically, the more complete the vegetation cover, the greater the amount of material displaced. This is because a dense root network will trap large amounts of unstable material before root cohesion is overcome.

This sort of mechanism appears to explain the cyclic landslides of Fiordland, New Zealand. Wright and Miller (1952) observed that on glacially oversteepened slopes in this part of the country, the vegetation cover induced a condition of instability on reaching a certain critical height and density. In this state, debris avalanches were readily triggered, producing a bare surface that in time became recolonised thus re-initiating a vegetation-induced cycle of instability. The length of time between the creation of a fresh erosional surface by mass movement and the development of a mature forest in the high rainfall region of Fiordland has been calculated to be at least 80 years (Mark et al. 1964).

'Tree levering' is another mechanism accompanying root wedging on steep, well-jointed, rock slopes. As a tree grows outwards from the slope it may act as a simple lever pivoting on the slope surface and transmitting stress to the root system. The weight of increasing biomass alone may be sufficient to dislodge loose blocks but the mechanism is most

effective in high winds when vibrations are transmitted
through the stem and root system to the rock face. If the
tree is actually thrown it is capable of dislodging blocks of
considerable size. Lutz (1960) has recorded the movement by
this process of blocks weighing 0.25 tonne through a distance
of up to 3.5 m.

Cleft-water pressure. Joints and other structural discontinuities which allow root penetration are also the means by
which other destabilising mechanisms can operate. The most
important of these results from the presence of water. In a
continuously jointed rock mass the level of water in the
joints can be extrapolated to demarcate a water table. The
presence exerted by cleft-water (as it is referred to)
corresponds in effect to the porewater pressure represented in
equation 3.1 and at any point is proportional to the distance
below the water table. Cleft-water pressure is exerted on the
joint surfaces reducing frictional resistance generated by the
overlying rock mass and, depending on the orientation of joint
surfaces, rock may be displaced outwards from the face.
As cleft-water is derived mainly from percolating rain-
water or meltwater, points of entry in the slope surface are
required. Terzaghi (1962) has pointed out that because of the
stress regime and the effect of weathering, joints capable of
receiving infiltration are likely to be more numerous and more
open towards the outside of the slope than they are further
away from the slope edge. As a consequence, the water table,
which in material of uniform permeability normally declines
with the slope, may at least temporarily increase in elevation
towards the outside of the slope. This increase in water
pressure in the outer zone of the slope material is enhanced
if joint outlets of the toe are sealed. Reduced permeability
in joint outlets can result from the formation of ice, accumulation of weathering products, artificial coverings such as
concrete skins and the exposure of compact joints through
recent excavation. Landslides triggered by cleft-water
pressure are indicated by a toe failure or 'blowout' at the
foot of the slope.

Ice wedging. Lateral pressure within joint systems may also
result from the increase in volume of substances occupying the
voids. For example, water at 0°C increases 9% in specific
volume on freezing under normal atmospheric conditions. The
pressure exerted by such unconfined freezing of water in the
form of needle ice (pipkrake) accumulating on the ground surface is capable of lifting blocks of material weighing up to
9.5 kg (Pitty 1971). Taber's (1929) classic experiments
showed that with adequate water supplies pressure results not
only from change in volume but also from crystal growth. The
molecular cohesion of ice attracts water which nourishes the
crystal, causing it to extend and set up maximum pressure in

the direction of growth; that is, perpendicular to the cooling surface.

However, within a rock mass freezing is much more difficult. The confining pressure generated within rock voids as freezing commences reduces the temperature at which freezing can continue. Laboratory experiments show that, at a confining pressure of 50 000 kPa (500 kg/cm^2) for example, the temperature required for freezing water drops to about -4°C. Freezing cannot continue against such confining pressures unless the temperature continues to drop or the rock or ice is dislodged to increase void space and lower confining pressures. In nature, water in very small interstices remains in the liquid state even when supercooled to low temperatures. On the other hand, in wide joints open to the surface, water will be forced out as ice crystallises and the ice itself may be extruded since it behaves as a plastic at stresses in excess of 100 kPa. However, if partial freezing of void water is very rapid, remaining air cavities may implode setting up high stress shock waves which are theoretically capable of splitting rock (Hodder 1976). An even more important constraint on progressive freezing is the heat released when water freezes (latent heat of fusion). In poorly drained periglacial areas this mechanism is responsible for stabilising falling temperatures in the upper ground layer at around 0°C for one to four months in early winter and is known as the 'zero curtain' effect (Price 1972). Partial freezing will also tend to concentrate salts in solution and inhibit further freezing.

Theoretically, the pressures that could be exerted by confined freezing are enormous and easily sufficient to shatter massive rock slopes and set off deep-seated landslides. In nature, however, frost shattering (frost wedging) affects only the outermost zone of the slope (to a maximum depth of about 2 m and usually much less) because either sufficiently low temperatures cannot be achieved at greater depths, or in polar regions where they do occur, lack of water in its liquid state is the principal constraint. The geomechanical effects of ice pressure (cryostatic pressure) which in nature does not exceed 600 kPa (Connell and Tombs 1971), are thus most evident in periglacial regions (sub-polar regions or in mid-latitude alpine areas) where the number of freeze-thaw cycles is high and penetrate some distance below the surface. In the sub-alpine location of Como in the Colorado Front Range, Fahey (1973) recorded up to 65 freeze/thaw cycles within the top 1 cm of soil in the 12 months preceding July 1970. At a depth of 10 cm the number of freeze/thaw cycles dropped to 40 but most notably the geomechanical response recorded in the form of frost/heave cycles dropped to one. Sixty-four freeze/thaw cycles have been recorded close to sea level in northern Norway while 81 were recorded at an elevation of 500 m (Dahl 1966).

The geomechanical importance attributed to freeze-thaw cycles has been challenged by White (1976) who suggests that hydration fracturing may be a more important cause of rock breakdown in periglacial environments. Hydration fracturing results from fatigue set up by the expansion and contraction of adsorbed non-freezable water on mineral surfaces in response to oscillations in relative humidity. In fact, it is difficult to separate the effects of freezing and thawing of bulk (absorbed) water from hydration fracturing, because oscillations in relative humidity accompany temperature cycles. Laboratory experiments carried out by Fahey (1983) establish the effectiveness of hydration fracturing at temperatures both above and below zero. However, only about one quarter of the material shed during the experiments was attributable to hydration, the rest is thought to have been produced by freeze-thaw action.

Salt wedging. Water within a rock mass is seldom pure, containing substances dissolved from the rock or natural and induced pollutants taken up by atmospheric moisture. In both hot and cold arid areas and even in other particular local climates where evaporation or sublimation exceed precipitation, crystals other than ice may grow, setting up lateral pressure in a rock mass. A growing crystal of alum, for example, has been shown to exert a force of 400 kPa (Birot 1968). Crystals increase in volume by taking up molecules of the precipitate, by adsorbing water (hydration) and by thermal expansion.

Aluminous, ferruginous, siliceous and calcareous minerals have all been identified as being capable of crystal growth within rock joints but probably the most common substance operating in rock break-up is halite (common salt). Even the hard quartz diorite of McMurdo Sound, Antarctica, appears to be primarily weathered by the crystallisation of halite (Kelly and Zumberge 1961) although the scarcity of free water in Antarctica may indicate that the mechanism is a relict of a former climatic regime (Wellman and Wilson 1965).

Although Goudie (1974) and Goudie and Watson (1984) have shown that the process of crystallisation is a most effective form of salt weathering, Cooke and Smalley (1968) have demonstrated that, with the exception of $CaCO_3$, certain naturally occurring salts ($NaNO_3$, $NaCl$, KCl, $BaSO_4$) show a thermal expansion greater than that of granite and that this phenomenon may be important in weakening rock. However, for a 1% increase in volume, large temperature changes are required (54°C for NaCl) and even though rock faces are exposed to extreme temperatures, the low thermal conductivity of rock suggests that this process is unlikely to be very effective.

According to Winkler and Wilhelm (1970), expansion resulting from hydration of crystals can generate very high pressures. They have calculated, for example, that a pressure of 50 MPa can be exerted when thernardite (Na_2SO_4) is hydrated to mirabilite ($Na_2SO_4.10H_2O$) as a result of increasing relative humidity from 70% to 100% at a temperature of 20°C.

Increasingly it has been realised that it is unrealistic to treat weathering mechanisms in isolation. Fahey (1983), for example, has assessed the combined effect of freeze-thaw and hydration fracturing (see Ice Wedging) and Goudie (1974), Williams and Robinson (1981) and McGreevy (1982) have studied the effect of saline solutions on freeze-thaw processes in chalk, sandstone, and limestone respectively. Whereas the work on chalk and sandstone indicates that the presence of sodium chloride and sodium sulphate enhances frost shattering, McGreevy's studies suggested that a number of factors modify the role of salt solutions. It appears that although frost shattering can be enhanced when the supply of salt is frequent and plentiful, if the supply is limited and the amount of salt remains constant, then rock breakdown is inhibited.

A number of other factors control the effectiveness of saline solutions in rock breakdown. The type of salt, for example, influences the penetration ability of solutions and the amount of freezing point depression. The concentration of the solution also affects penetration ability as well as the amount of ice formation. In addition, the range and fluctuation of temperature produces measurable differences in rock breakdown.

Water wedging. The thermal expansion of water itself may also be effective in rock breakdown as it has a higher coefficient of expansion than have most rocks and, according to Winkler (1977), may be capable of exerting a stress of 25 MPa on heating from 10°C to 50°C in minor pore spaces.

In their work in hot desert regions, Ravina and Zaslavsky (1974) have suggested another mechanism for enlarging rock cracks. Water condensing on the crack surface at night sets up an electrical gradient in the water between adjacent rock surfaces. Pressure from the double layers on each surface will interact and disruptive swelling may occur. A similar process referred to as 'hydrofracturing' (Hradek 1977) occurs as capillary films of water are forced to the distal tips of cracks by pressure of surface ice formation.

The mechanism of hydration fracturing which is also dependent on oscillations in temperature and relative humidity has already been discussed in conjunction with freeze-thaw activity (see Ice Wedging).

Clay swelling. Substances other than precipitates, water and ice may occupy joints and voids within a rock mass and be responsible for the generation of lateral pressure. The most common of these is clay. It can occur within joints either as a result of direct chemical weathering of the joint surface or from deposition by gravitational water. Common rock forming minerals such as feldspar react with water to form hydrous alumino-silicates (clay). By the chemical process of hydration and hydrolysis, complete or dissociated water molecules become combined with the primary mineral, altering it and causing an increase in volume. A much more rapid increase in volume, however, occurs within the clay itself as layer water is adsorbed and absorbed by negative charges on the surface of the clay platelet. Given unlimited water, the rate at which swelling occurs depends on the initial moisture content. Because the platelets in 'dry' clay are tightly packed together, porosity is low and consequently initial entry of water is difficult. In any one type of clay, the degree of volume change as a result of swelling also depends on the initial water content as well as on the overburden or confining pressure.

Figure 3.10 illustrates an idealised relationship between the initial moisture content and the amount of volume change that can take place under different overburden pressures. The initial moisture content scale is shown originating at the 'shrinkage limit'; that is, the moisture content at which there is no further decrease in volume with further evaporation of porewater (see Section 3.3.11 MATERIAL AND STRUCTURE - Soils). In this example, an unconfined clay with a 10% moisture content at shrinkage limit can swell 30% in volume. The same unconfined clay, if already at a moisture content of 20%, can only swell by 10%. As the curves flatten out at higher moisture contents (Kassiff et al. 1969), a given increment of moisture change will be accompanied by less swelling in 'wet' clays than in drier clays. This is because, in the first stage of swelling, absorbed water entering between the clay platelets reduces tensions more effectively than does free water that is drawn in by capillarity in later stages. Figure 3.10 also shows that at a given moisture content, the amount of swelling is reduced by an amount approximately equivalent to the logarithm of the overburden pressure. Thus with equivalent initial moisture conditions a clay which can expand 10% when unconfined can expand only 2% when subject to an overburden pressure of 1 kPa.

In this particular example the greatest load that swelling could shift would be about 10 kPa and this would occur only if the sample were very dry; that is, close to its shrinkage limit. The overburden load which is just sufficient to prevent swelling is referred to as the 'swell pressure'. The swell pressure applied to any swollen clay will consolidate the material (given adequate drainage) until its void ratio is that attained at the shrinkage limit. Means and Parcher (1963) note that theoretically the swell pressure will be equal to the capillary tension at shrinkage limit. Thus tensions at shrinkage limit can give an indication of the maximum stress than can be exerted by swelling.

Both the amount of clay in a sample (measured by its

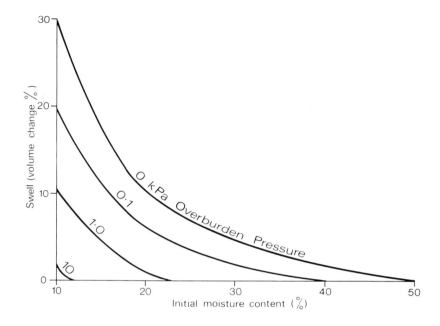

Figure 3.10 Clay swelling related to initial moisture content and overburden pressure for a hypothetical clay (generalised from Kassiff et al. 1969)

colloidal content) and the type of clay (as indicated by its physical properties) determine the degree of swell pressure that can be generated within a soil **(Table 3.2).**

Table 3.2 Swelling and Shrinkage Potential of Clays Related to Index Properties (adapted from Holtz and Gibbs, 1956, by Lee et al. 1983)

Colloid content (%)	Plasticity index (%)	Shrinkage limit (%)	Swell volume change (%)	Degree of swelling
>28	>35	<10	>30	very high
20-13	25-40	7-10	20-30	high
13-23	15-30	10-15	10-30	medium
<15	<18	>15	<10	low

The physical index properties (see Section 3.3.11 MATERIAL AND STRUCTURE - Soils) in turn are largely explained by the mineralogy of the clay. **Figure 3.11** shows that the degree of plasticity of a soil depends not only on the amount of clay but also on the type of clay. The degree of swelling in turn depends on plasticity. One of the clay minerals shown, montmorillonite, belongs to a group of expanding lattice minerals, some of which can swell to several times their initial volume.

The swelling and shrinking of material on joint planes inclined towards the outside of the slope may progressively open joints by a form of joint-block creep induced by direct contact pressures. Swelling may also have an indirect effect by pressurising entrapped air at the open ends of cracks causing 'air breakage' of rocks and propagation of cracks (Taylor and Spears 1970).

The agents responsible for lateral pressure discussed in this section are those which at least theoretically could work in existing joints and other openings to displace rock towards the outside of the slope. Although some of the processes are thought to be capable of actually propagating cracks in rocks of low tensile strength, their greatest effect will be in the outer zone of the slope where intersecting joints already exist.

Apart from the action of cleft-water pressure, these agents must be considered to be preparatory factors in slope destabilisation either weakening the rock mass or providing void space for the entry and accumulation of storm water. They belong to one end of the weathering spectrum which starts almost from the moment that rock is separated from its original petrogenic environment. Whereas Whalley et al. (1982) have examined the mechanisms for initial microfracturing in the early stages of weathering, Stewart (1964) has assessed the net effect of prolonged weathering. He noted, for example, that weathering of crystalline rocks can increase porosity from initial values of less than 0.05% to between 30% and 60%.

There are many problems in assessing the individual effectiveness of various weathering agents in nature. The main difficulty lies in imposing appropriate boundary conditions which are applicable to the field situations upon the results of theoretically-derived or laboratory-demonstrated processes. Little is known about the regime of so many of the important controlling factors within the naturally-occurring body of rock, such as humidity, temperature, pressure and salinity. As rock properties are so variable and multiple weathering processes can occur simultaneously, assessing the effect of one process becomes exceedingly difficult.

On steep slopes, agents that operate even intermittently will produce a cumulative opening of joints. This occurs not only as a result of gravitational forces but also from a pro-

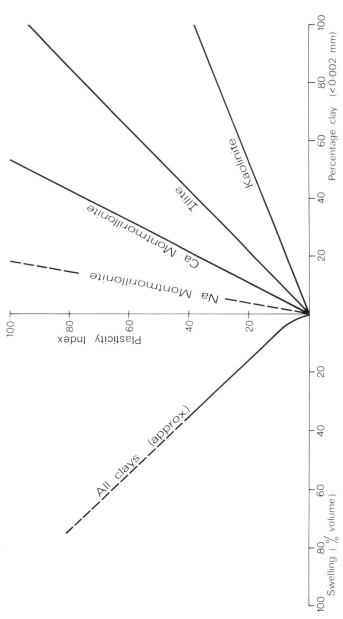

Figure 3.11   Relationship between the amount and type of clay and its plasticity and swelling potential (adapted from Skempton 1953 and Holtz and Gibbs 1956)

cess of 'debris chocking'. Small rock fragments and other material transported into joints play an extremely important complementary role in the weathering process, by holding rock surfaces apart. This process and other factors such as the vegetation effect are difficult to reproduce in laboratory experiments; confirmation of the potency of a number of joint processes must therefore await adequate field experimentation.

3.3.8 SEISMICITY AND OTHER VIBRATIONS

Vibrations from traffic, machinery, explosives and wind rarely trigger or accelerate slope movement. Shaking from these sources is effective only in areas of extreme inherent instability close to the site of energy release. In southeast Alaska, for example, Vandre and Swanston (1977) noted that the two large debris avalanches triggered by quarry blasting on Mitkof Island occurred on steep slopes when soils were already saturated. In the course of their analysis, they made the observation that, under certain circumstances, blasting could produce accelerations in soil as high as those experienced in the underlying rock - a response which is not normally predicted and which should be kept in mind during site investigations.

Benson (1940) identified an increase in the weight and frequency of rail traffic as the cause of accelerated slump movement in eastern Otago, New Zealand. Records of railway tunnel alignment for a 58 year period from 1878 showed an 80% increase in the rate of annual displacement and cessation of movement within two years of the tunnel being abandoned. The tunnel traversed a compound slump in sandstone which was underlain by plastically-deforming mudstone. Unlike many landslides in the area the direction of movement is opposed to the dip direction of the strata but lateral support has been removed by the erosion of a valley into soft mudstone.

Of far wider significance than any of the other sources of vibration are earthquakes (**Figure 3.12**). Keefer (1984) in a comprehensive study has identified 14 types of landslide (including submarine landslides) associated with earthquake activity: the most common are rock falls, rock slides and disrupted soil slides. His observation, that certain slope movements within soil are among those slides initiated by the weakest effective shaking (Modified Mercalli Scale VI), should dispel the view that soil failures rarely result from earthquakes (Hawley 1984). Although much of the soil material in slope movements caused by earthquakes has had a high water content at the time of shock (for example, the Bootlegger Cove clay of Anchorage, Alaska; Hansen 1966) some has been dry, such as the Kansu loess involved in the devastating landslides of China in 1920 (Coates 1977).

Earthquakes reduce stability by imparting both a shearing stress and a reduction in resistance to slope material.

Figure 3.12 Rock falls and debris slides resulting from the 1929 and 1968 earthquakes, Matiri Tops northwest Nelson, New Zealand (Photo: Professor D W McKenzie)

Earthquake wave propagation is thought to have three principal effects:

1. the direct mechanical effect of horizontal accelera-tion which, at high shaking intensity, may exceed acceleration due to gravity. This provides a tem-porary increment to shearing stress which is suf-ficient on occasions to trigger landslides;

2. cyclic loading in clays, sands, and silts with weak inter-particle bonding. In saturated material, seismic loading shifts the weight of particles from its granular support onto the porewater, thereby increasing interstitial pressure, buoying up the mass and causing liquefaction. Sensitive clays (see Section 3.3.11 MATERIAL AND STRUCTURE) are par-ticularly susceptible to earthquake shaking and other vibrations;

3. reduction of intergranular bonding afforded by cohe-sion and internal friction, by sudden shock, irrespective of the degree of saturation. This lowers the strength of material towards its residual value. The effect is similar to that experienced by a brick building when shaking separates the bonds between mortar and bricks. Although this response may not immediately initiate movement, it serves to make the slope susceptible to future triggering activity.

3.3.9   CHANGES IN WATER CONTENT

Changes in water content can quickly affect the stability of slope material and have been responsible for triggering, re-initiating and accelerating more landslides than has any other factor. In nearly all cases an increase in water content decreases stability in one or more of the following ways:

1. increasing interstitial porewater pressure. The way in which this factor operates is illustrated in equation 3.1 and its accompanying explanation. Positive porewater pressures which reduce resistance can be developed within the phreatic zone or within groundwater zones perched above a relatively imper-meable substrate (see Section 3.3.10 WEATHERING AND ACCUMULATION);

2. developing cleft-water pressure within joints, voids and fissures (see Section 3.3.7 JOINT PROCESSES - Cleft-water pressure). This has a similar effect on resistance to that produced by interstitial porewater pressure (equation 3.1). The effect of water pressure on the value of strength parameters (c and $\emptyset$) required to maintain stability is illustrated in **Figure 3.13.** The position of line (d) in this figure

79

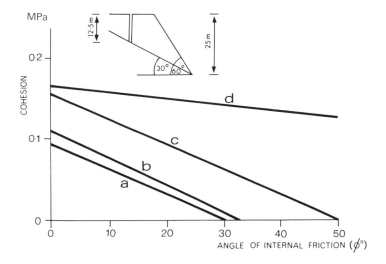

Figure 3.13 The magnitude of shear strength parameters required to maintain stability under various conditions of two-dimensional plane failure (Hoek and Londe 1974):
a. dry slope - no tension crack
b. dry slope - tension crack
c. water-filled tension crack
d. water-filled tension crack and water pressure on failure plane.

indicates the overriding influence of both cleft-water and interstitial water pressure on reducing the effect of the angle of internal friction. In this example it is unlikely that material, even with extremely high angles of internal friction, could withstand the saturated conditions represented by line (d), unless it also possessed high cohesion. As apparent cohesion afforded by clay particles would soon diminish under these conditions, stability could be maintained only by true cohesion.

The other lines in Figure 3.13, c, b and a, represent progressively more stable conditions. The greater stability of (c) compared to (d) is due to the absence of continuous groundwater above the slide plane. The improved stability represented by (b) results from the absence of water in the tension crack and the most stable situation, represented by (a), is attained by an intact slope under dry conditions. While it is clear that stability is reduced by the presence of a tension crack, in this example, cleft water and porewater have an even greater influence;

3. developing seepage pressure where a drag stress is set up in the direction of water percolation thus contributing to shear stress. Seepage pressure may also lead to gradual subterranean erosion and removal of underlying support (see Section 3.3.6 REMOVAL OF UNDERLYING SUPPORT);

4. imparting 'lubrication'. Terzaghi (1950) emphasised that by itself, water generally acts as an anti-lubricant with respect to the frictional coefficient of most rock-forming minerals. However, by converting joint-fillings and thin, clay-rich, inter-bedded layers into slurries, a form of 'lubricant' is developed which can facilitate sliding of overlying material;

5. increasing weight. The effect of increase in weight provided by water is difficult to separate by analysis from the effects of porewater pressure and reduction in cohesion and consequently it has some-times been ignored. As noted in Section 3.3.3 VARIATION IN WEIGHT, increase of weight can only instigate failure in cohesive slope material or in certain precarious circumstances where the stability of a fractured material is being maintained solely by root cohesion. Slopes consisting of clay, par-ticularly the expanding-lattice variety, are the most

susceptible to triggering by this mechanism. Such
material maintains its strength largely by cohesion,
can take up large amounts of water (sometimes
doubling unit weight of the material), and commonly
exhibits desiccation cracking which facilitates the
entry of water;

6. decreasing cohesion (apparent cohesion). In soils,
cohesion (equation 3.1) results mainly from capillary
and electro-molecular forces which are reduced as the
amount of interstitial water increases. In weathered
granite saprolite, for example, Lumb (1965) has noted
that apparent cohesion can be as high as 200 kPa but
that it will drop to zero when the material is
saturated. When soil moisture is sufficient to
neutralise cohesive forces any additional water,
given sufficient pore space, will percolate through
the system as gravitational water. When drainage or
pore space become limiting factors, additional water
will instead form a continuum of groundwater thus
exerting positive porewater pressure which along with
reduced cohesion will further destabilise the slope.
In some fine soils, water content may exceed the
liquid limit (the water content at which a remoulded
sample behaves as a liquid: see Section 3.3.11
MATERIAL AND STRUCTURE - Soils), without causing
failure (Vandre and Swanston 1977; Owen 1981). In
such cases, cohesion is likely to be negligible and
strength is afforded only by the frictional contact
of coarse particles. When these contacts are broken
by positive porewater pressures or other stresses,
the material moves downslope as a fluid. In very
fine, shallow soil without a frictional component and
with low liquid limits (Owen 1981), spontaneous
liquefaction may occur unless the soil is constrained
by stronger neighbouring material or a strong,
coherent root mat. Apparent cohesion may also be
destroyed by reduction in water content. For
example, triggering of a rock slide in Cerro
Condor-Seneca, Peru, which occurred during an abnor-
mally dry season, has been attributed to desiccation
and cracking of clay joint-fillings in granodiorite
(Snow 1964). The process of 'dry ravelling' whereby
particulate matter is shed from exposed faces during
dry weather is also a response to the same mechanism.

For changes of water content to occur there must be a
variable source of water and a capability for infiltration to
exceed drainage. Most commonly, variations in precipitation,
snowmelt and evapotranspiration are ultimately responsible for
changes in water content of slope material. However, the

supply of water may be concentrated onto or within the slope by uneven topography, differences in permeability or by artificial means. Irrigation, faulty or careless stormwater disposal, and leaking pipes and reservoirs have all at one time or another contributed to the initiation of landslides.

3.3.10 WEATHERING AND ACCUMULATION

Weathering, in its broadest sense, involves all those surficial processes (physical, chemical, and biological) which slowly alter the nature of material without actually involving its erosion and transport. Under this definition, the actions of some of the stability factors already discussed, particularly the joint processes, constitute a form of weathering.

The weathering process, given time, can convert rocks of very high shear strength into soil of very low shear strength. The degree and rate of this change depends not only on the original mineralogy but also on the nature of the climatic and biological environment. In tropical countries, for example, weathering products formed from hard igneous rocks may accumulate to depths in excess of 40 m on gentle parts of the topography. Weathering changes such as these are accompanied by a systematic decrease in the critical angle for slope stability (Durgin 1977).

Surficial material may also progressively undergo a change in properties as a result of the incorporation and accumulation of exogenous material such as loess, volcanic ash and material deposited from further upslope. In addition, changes can be brought about by the removal of particulate, colloidal and cementing substances through the processes of eluviation and solution (Ruxton 1958).

An appreciation of weathering and other surficial changes is important for understanding the development of instability and for assisting in the identification of unstable sites.

In the process of its conversion to soil, bedrock initially experiences a loss of cohesion as joints develop and open (Terzaghi 1962). Close to the surface where joints intersect, cohesion is lost completely and the affected rock mass obtains its strength entirely from the interlocking and planar friction of joint-blocks. Where slope, vegetation, and climatic conditions favour the retention of jointed rock, weathering continues in situ, exploiting the increased surface area afforded by jointing (see Section 3.3.7 JOINT PROCESSES).

In any one bio-climatic environment, properties of the accumulating weathering products are a function of time. Young weathering products have proportionally more rock fragments and primary minerals and less clay than do older products. As weathering continues, the proportions change in favour of clay. Accompanying this progression, there is a change in void ratio, a reduction in internal friction and

permeability and re-establishment of a cohesive potential.
The changes brought about in these and other properties
have been used as the basis for classifying the degree of
weathering. Comparative measurements indicating the degree of
weathering can be made using the following properties. (The
+ sign indicates that the value increases with the degree of
weathering and the - sign indicates a decrease):

1. ratio of secondary minerals (usually clay) to primary
   minerals (+);
2. mean particle size (-);
3. plasticity (+);
4. appearance of index minerals from the mineral stability series;
5. depth of weathering profile (+);
6. discolouration (+);
7. preservation of original rock fabric (-);
8. thickness of weathering rind (+);
9. Schmidt rebound hardness number (-);
10. void ratio or porosity (+);
11. dry density (-);
12. permeability (+);
13. distance (spacing) between joints (+);
14. openness of joints (+);
15. amount of joint filling.

Table 3.3 shows a weathering classification applicable to
hard rock. Fookes and Horswill's (1970) original scheme, from
which it is derived, also provides a classification of
weathering in soft rock. The classification scheme illustrated has been used by a number of workers to correlate
weathering grade with properties of significance to slope stability **(Table 3.4)**. Pender (1971), however, found that
weathering-related variation in void ratio alone provided a
satisfactory indication of shear strength parameters and other
mechanical properties.
Void ratio not only responds to weathering changes but
also to accumulation and incorporation of exogenous material.
Fine sediments brought onto the slope may be washed into the
interstices of the underlying material thereby reducing the
void ratio and affecting inherent stability conditions. On
formerly stable scree slopes, for example, this process
appears to reduce permeability significantly, facilitate the
development of positive porewater pressures and consequently
lead to the initiation of shallow debris slides during wet
conditions.
Measurement of the strength parameters for partially
weathered rock has proved to be a problem because the lowest
strength (particularly in rock of grades I-III) often occurs
along variously disposed joint-surfaces which are difficult to
test adequately with conventional shear tests. Specially
developed direct-shear equipment (Hoek 1970; Martin and Millar

Table 3.3 **Classification Scheme for Weathered Greywacke**
(after Fookes and Horswill (1970), modified by
Martin and Millar (1974))

Term	Grade	Description
True residual soil	VI	Original rock fabric completely destroyed. Rock completely changed to soil, generally light or yellow-brown sandy clay.
Completely weathered	V	Original rock structure completely weathered - crushable to light brown sandy silts under finger pressure. Original rock fabric still visible, with joint patterns marked by iron or black manganese dioxide stains.
Highly weathered	IV	Original rock structure retained but generally weathered to light brown colour right through. Most of material can be crushed to silt and sand sizes under finger pressure, but harder lumps remain. Rock structure generally open and closely jointed.
Moderately weathered	III	Original rock structure retained. Brown weathering extends part way through rock fragments, leaving grey unweathered central core. Rock structure tighter. Rock fragments easily broken with light hammer blow.
Slightly weathered	II	Hard jointed rock. Brown colour extends inwards a short distance on joint planes. Interior has colour and texture of unweathered greywacke. Separate pieces require moderate hammer blow to break.
Fresh rock	I	Unweathered greywacke. Shows no discolouration, loss of strength or any other effects due to weathering.

Table 3.4 Properties of Weathered Greywacke

Weathering Grade	Schmidt Hardness	Dry Bulk Density (kg/m³)	Porosity (%)	ϕ (degrees)	Unconfined compressive strength (MPa)
VI	0	1600	64	15–32	0.5
V	0	1500	>23	35–15	0.8
IV	0–10	2020–2540	17–25	35	1.0
III	10–20	1930–2420	9–20	35–45	15–20
II	15–25	2320–2450	7–13	35–45	20–35
I	25–40	2540–2570	<9	35–45	35–80

Note: The values in this table are derived from a number of different localities and have been taken from the following references: Pender (1971), Martin and Millar (1974), Riddolls and Perrin (1975) and McConchie (1977)

1974) can, however, be used to test joint strength. These tests generally indicate that, contrary to Coulomb's law (equation 3.1), the relationship between normal stress (σ) and the angle of internal friction ($\mathscr{B}$) cannot be represented by a straight line under all conditions. According to Patton (1966), at low normal stresses, resistance is being generated by both planar friction of the joint-surface and the force required to produce dilatancy and overriding of asperities along the joint surface. When this occurs the angle of internal friction generated is equal to ($\mathscr{B}_u + \mathscr{B}_i$); that is, the peak angle of planar friction plus the inclination asperities make with the shear surface. At high normal stresses, on the other hand, the asperities are sheared-off, resistance per increment of normal stress is reduced and the angle of internal friction is therefore lower, corresponding approximately to the residual angle of friction ($\mathscr{B}_r$). Asperities become softer and less resistant as weathering advances and rock of grade IV or above usually shows no increment or resistance attributable to roughness of the joint surfaces.

The variation in behaviour of weathered and jointed rock dictates that care must be taken to ensure that strength testing is carried out with samples of the appropriate weathering grade subject to a loading applicable to the slope under consideration.

Cohesion developed in completely weathered material and residual soil is much less effective than that of the parent rock and is subject to rapid changes in magnitude. Saturation, for example, can temporarily reduce or even eliminate soil cohesion (Lumb 1962) and slow mass movement, such as soil creep, can also destroy cohesion as well as reduce internal friction to its residual value. Consequently, some long-term, limiting equilibrium analyses of hillslope soils incorporate the residual angle of internal friction as the only effective strength parameter operating within the slope material (Carson and Petley 1970; O'Loughlin 1974).

Mass movement and other erosional processes may remove weathering products from their place of origin at any stage of the weathering process. However, one of the most important stages with respect to slope stability is represented by the formation of a discrete regolith, separated from the underlying parent-material or bedrock by a distinct, weathering-front surface. The strength and hydrological properties on either side of this surface are usually very different (Figure 3.14). A similar surface, representing a major-discontinuity in soil properties, may be present where slope or air-fall deposits overlie bedrock. These surfaces generally serve as shear planes for shallow soil slides (Barata 1969; Durgin 1977; Crozier et al. 1981 and 1982) (Figure 3.15). Their significances, in some cases, can be attributed to a loss in strength owing to the removal of

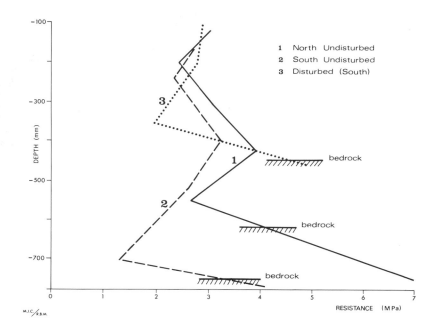

Figure 3.14 **Penetration resistance with depth for a clayey-silt regolith at three localities.** Note the reduction in resistance immediately above the bedrock surface.

material by seepage of water along the surface. More importantly, they increase the likelihood of perched water tables forming above the bedrock. Both these mechanisms require an abrupt reduction in permeability at depth. Pedological processes may also produce relatively impermeable horizons within the soil and thus allow similar mechanisms of instability to operate.

The degree of weathering is also fundamental in determining the type of landslide that takes place - a fact that is recognised by many classifications. One example of its influence on the type of landslide produced is shown in **Table 3.5.**

Durgin's (1977) work in granitic material illustrated in Table 3.5 indicates the importance of the bedrock surface and joints in dictating the position of the slide surface. Even in completely weathered material, he observed that residual

Figure 3.15   Mass movement erosion of volcanic ash deposits overlying greywacke basement rocks, Kaweka Range, Hawke's Bay, New Zealand

joints containing precipitates of manganese and iron had a
shear strength one-half to two-thirds that of the surrounding
saprolite. In weathered greywacke, however, Martin and Millar
(1974) have found that, at weathering grade IV or greater
(Table 3.3), the strength of joints more or less equals that
of the intact rock and consequently joints are less influential in controlling the position of the shear plane.

Table 3.5   **Influence of Weathering on Landslide Type in
Granitic Terrain** (after Durgin 1977)

Weathering Stage	Weathering Products	Landslide Type	Slide Surface
Fresh Rock	<15%	rockfalls, rock slides, block-glides	joint-surfaces
Corestone	15%-85%	rockfalls, rock avalanches, rolling rocks	sheeting joints, bedrock surface
Decomposed Granitoid	85%-100%	debris flows, debris avalanches, debris slides	bedrock surface
Saprolite	100% + laterite	rotational slides	residual joints

3.3.11   MATERIAL AND STRUCTURE

Slope material and its structure are important because they
determine the tolerance of the ground to all other destabilising factors. Their combined influence is represented by
the inherent strength of material. The greater the inherent
strength, the more difficult it is for the other destabilising
factors to produce failure. In engineering terms, this means
that material with greater inherent strength can more safely
withstand steeper and higher cut slopes than can weaker
material.

If this is true, what then do steep and high slopes
represent in their natural environment? Do they indicate
strong material and thus stable conditions or, as is often the
case, should such slopes be considered the most likely sites
for landslides? This is an important question because slope
form is one of the most commonly used parameters in regional
stability assessments (Cooke and Doornkamp 1974). The relative importance of stability factors has already been covered
in previous sections (Section 3.1 THE CONCEPT and 3.2

PRINCIPLES OF CAUSATION), and that discussion indicated that one factor alone, in this case slope form, cannot be used to judge stability. Nevertheless, an examination of a world-wide distribution of steep slopes will indicate that most of them are composed of strong material. However, some of these will have a low margin of stability because of factors such as climate, susceptibility to undercutting or because they are already close to their limiting angle for stability. Conversely, tectonic activity can maintain high relief in relatively weak material but this will also be subject to continual mass movement as it undergoes adjustment.

Gregory and Brown (1966) and Selby (1982a) have suggested that in any one location, the characteristic slope angles (Young 1961) are a function of rock conditions. This relationship however, will only apply where the terrain is in equilibrium with environmental conditions, where all slopes have been subject to the same processes for a similar duration and where slopes are erosional as opposed to constructional. The achievement of the conditions would require a long period free from tectonic or climatic disturbance and sufficient time for unstable material to be shed from the slopes. The rate of slope weakening as a result of erosional, overburden release and weathering may mean that in some rock types, slope adjustment by landsliding is an on-going and perhaps cyclic process, separated by long periods of apparent stability. Slope form itself, therefore, is not a universally reliable indicator of inherent strength.

As Skempton (1953) observed, it is often necessary to refer to two types of slope stability which reflect both inherent strength and the prevailing process. These are stability in respect of shallow landslides and stability in respect of deep-seated landslides (Figure 3.5). The natural slope, as described in Section 3.3.10 WEATHERING AND ACCUMULATION, is generally composed of bedrock and overlying regolith. Although both materials may possess cohesion, that within the regolith is apparent cohesion only and can be eliminated under saturated conditions, leaving residual internal friction as the only effective strength parameter. Consequently, slope angle, and not height, will be the main landform control of regolith stability. Limiting equilibrium, infinite slope analysis applied to this kind of situation (Skempton and DeLory 1957) indicates that the critical slope angle for stability is approximately equal to half the residual angle of internal friction. Strength of the underlying bedrock, on the other hand, is much greater than that of the regolith and depends on cohesion as well as internal friction. Whereas shallow landslides, given sufficient slope angle, can occur with little relief, deep-seated landslides involving bedrock require a critical slope height to be surpassed before they can occur. Thus shallow instability can be achieved on stable bedrock slopes (see also Section 4.2 FLUVIAL DOWNCUTTING).

Classifying material. From the foregoing examples it is evident that slope stability conditions vary significantly with the type of material involved. It is therefore useful to classify the full range of natural materials in a way which will reflect their inherent strength. Unfortunately, a conventional Coulomb-Terzaghi determination of shear strength on a sample of intact material does not always reflect the true inherent strength of slope material. Inherent strength of a slope material is a function of both material properties and structure. Theoretically, inherent strength can be represented by internal friction and cohesion, which are properties of the material. However, structural factors determine the extent to which these strength parameters can be mobilised, as well determining the location, within the slope, of material with the least resistance. The degree to which conventional measures of strength truthfully represent inherent strength should be kept in mind when reviewing how earth scientists and engineers have classified slope material.

Rocks and material derived from them have traditionally served as the object of study for geologists and geomorphologists and consequently they have been responsible for their original classification and nomenclature. As a result, natural materials have been classified principally by mode of origin, mode of transport, mineralogy and constituent particle-size, rather than by any direct consideration of geomechanical properties. The stringent demands of modern engineering, however, have meant that engineers have been required to adopt their own classification of material. In between the engineers and geologists lies a relatively newly formed group, the engineering geologists, who aim at characterising the established geological rock types in geotechnical terms and whose nomenclature is often a mixture of geological and engineering terms.

To the engineer, natural materials can be divided into two major groups on the basis of strength and behaviour: rock and soil (**Figure 3.16**). Rock is hard and rigid, may require blasting during excavation, and is not greatly affected by immersion in water. Soil, on the other hand, can be 'separated by such gentle means as agitation in water' (Terzaghi and Peck 1967). In other words, rock possesses 'true cohesion' and soil is either cohesionless or has only 'apparent cohesion'. True cohesion originates from cementing and fusion of mineral particles whereas apparent cohesion results from electro-molecular forces generated between fine soil particles in the presence of a thin layer of water (see Section 3.3.9 CHANGES IN WATER CONTENT). As discussed in the following section (Soils), soil may be classed as being either 'cohesionless' or 'cohesive' (possessing apparent cohesion). As indicated in Figure 3.16, within the definition of engineering soil, there are a number of different types of material which the geologist would not consider to be soil.

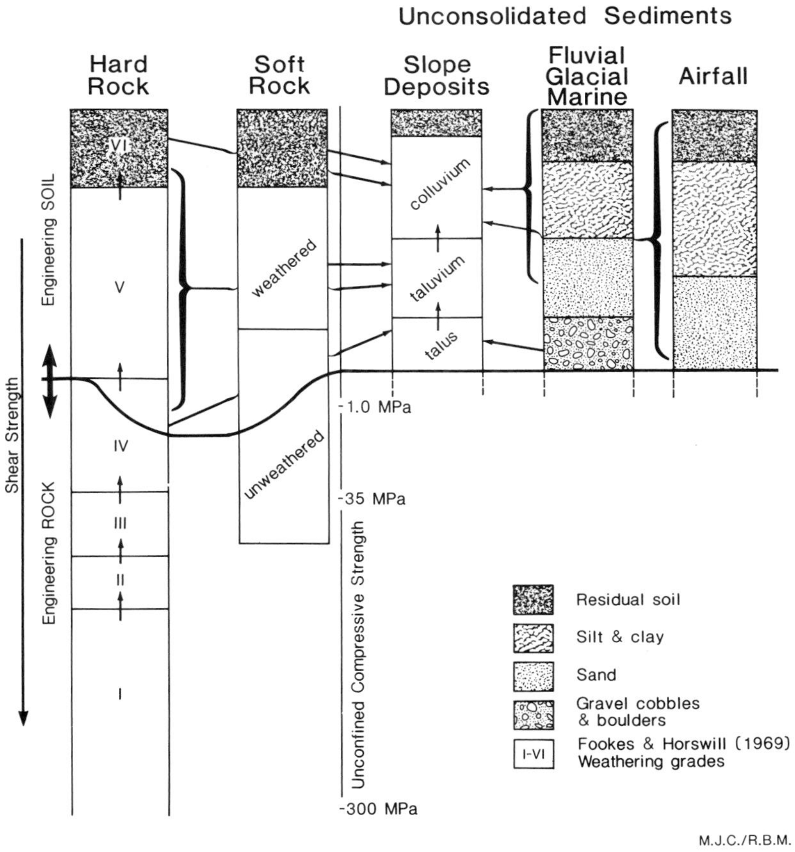

Figure 3.16   Types, relative strength and derivation of
                slope material

These materials (such as talus, colluvium, volcanic ash, river gravel and some soft rock, to name a few) have properties dominated by their mode of transport and deposition and have not been sufficiently altered by pedological processes to constitute 'true soil'. The arrows in Figure 3.16 also indicate that successively weaker materials may be formed by either weathering in situ or by erosion and transport.

The strength of both rock and soil is generally determined by the amount of shear, compressive, torsional, or tensile stress that can be withstood before the material fails. Of these four measures, shear strength (equation 3.1) most closely represents the kind of stresses experienced in slope failure. However, its measurement can be difficult and expensive and consequently other methods are sometimes used to provide a relative measure of shear strength. The most common of these is uniaxial (unconfined) compressive strength (q_u) and is related to shear strength parameters by the equation:

$$q_u = 2c.\tan (45 + \theta/2) \qquad \qquad (3.10)$$

Further simplification in the representation of strength can be seen in the attempts to relate q_u to simple field tests. **Table 3.6** represents one such approach for intact cohesive material. Whereas strength can be seen to form a continuum between the strongest and weakest materials, behaviour (that is, response of cohesion to water) serves to distinguish the two main groups: soil and rock. Behaviour under shear stress is also used in some classifications to subdivide rock strength categories. For example, Coates (1964) uses pre-failure deformation characteristics to classify rocks as either elastic or viscous and makes a further subdivision, by type of failure, into brittle or plastic rocks.

Soils. Geomechanical properties are also used to classify soils into three main groups which correspond with certain established particle-size grades (**Table 3.7**):
1. frictional soils; behaviour related to particle-size, shape and compaction; non-cohesive; dominated by sands, gravel and coarser fractions;
2. cohesive soils: behaviour related to both particle-size and mineralogy; exhibit apparent cohesion -
   (a) silts: material which has the properties of a silt although may not consist totally of the silt-grade fraction; cohesive but not plastic; undergoes dilatancy on disturbance; tends to liquefy on shaking; cracks and fractures more readily than does clay;
   (b) clays: material which has the properties of clay but may have less than 50% of particles in the clay grade fraction; cohesive and plastic;

Table 3.6 Approximate Strength Classification of Cohesive Soil and Intact Rock (Selby 1982c)

Description	Uniaxial compressive strength, MPa	Point-load strength, I_s (50), MPa	Schmidt Hammer, N-type, 'R'	Characteristic rocks
VERY SOFT SOIL – easily moulded with fingers, shows distinct heel marks	<0.04			
SOFT SOIL – moulds with strong pressure from fingers, shows faint heel marks	0.04–0.08			
FIRM SOIL – very difficult to mould with fingers, indented with finger nail, difficult to cut with hand spade	0.08–0.15			
STIFF SOIL – cannot be moulded with fingers, cannot be cut with hand spade, requires hand picking for excavation	0.15–0.60			
VERY STIFF SOIL – very tough, difficult to move with hand pick, pneumatic spade required for excavation	0.60–1.0	0.02–0.04		
VERY WEAK ROCK – crumbles under sharp blows with geological pick point, can be cut with pocket knife	1–25	0.04–1.0	10–35	Weathered and weakly compacted sedimentary rocks – chalk, rock salt
WEAK ROCK – shallow cuts or scraping with pocket knife with difficulty, pick point indents deeply with firm blow	25–50	1.0–1.5	35–40	Weakly cemented sedimentary rocks – coal, siltstone, also schist
MODERATELY STRONG ROCK – knife cannot be used to scrape or peel surface, shallow indentations under firm blow from pick point	50–100	1.5–4.0	40–50	Competent sedimentary rocks – sandstone, shale, slate
STRONG ROCK – hand-held sample breaks with one firm blow from hammer end of geological pick	100–200	4.0–10.0	50–60	Competent igneous and metamorphic rocks – marble, granite, gneiss
VERY STRONG ROCK – requires many blows from geological pick to break intact sample	>200	>10	>60	Dense fine-grained igneous and metamorphic rocks – quartzite, dolerite, gabbro, basalt

Data from: Deere and Miller (1966), Piteau (1971), Robertson (1971), Broch and Franklin (1972), Hoek and Bray (1977)

little dilatancy;
3. organic soils: high void ratios; high potential water contents; low dry density; permanent shrinkage; very compressible; dominated by decayed plant material.

Table 3.7   Particle-size Classification

Grade		Particle size range (diameter mm)
Clay		<0.002
Silt		0.002-0.06
Sand	- fine	0.06 -0.2
	- medium	0.2 -0.6
	- coarse	0.6 -2.0
Gravel	- fine	2.0 -20.0
	- coarse	20.0 -60.0
Cobbles		60.0 -300
Boulders		>300

The fundamental differences among these three soil groups are recognised by a number of geomechanical soil classifications. One of the most commonly used is the Unified Soil Classification which was first developed by Casagrande (1948) and subsequently adopted by the American Society for Testing and Materials in 1969 (ASTM, D-2487). Procedures for applying the system can also be found in many text books (for example, Lee et al. 1983).

The Unified Soil Classification relies principally on particle-size, particle sorting and Atterberg limits, particularly plasticity, to define 15 soil types. The use of these parameters allows each of the soil types to reflect differences in permeability, dry strength, compressibility and volume change as well as general workability for construction purposes.

The Atterberg limits (already referred to in Section 3.3.7 JOINT PROCESSES - Clay swelling) are of greater value than is particle-size distribution for characteristing cohesive soils because they reflect both particle-size and mineralogy. They are defined as follows:
1. liquid limit (W_L): the minimum moisture content at which a soil will flow under its own weight. It

defines the boundary condition between the liquid and plastic states;

2. plastic limit (W_P): the minimum moisture content at which the soil can be rolled to a thread up to approximately 3 mm in diameter without breaking. It defines the boundary conditions between the plastic and semi-solid states;

3. shrinkage limit (W_S): the moisture content at which further loss of moisture does not cause shrinkage. It defines the boundary conditions between the semi-solid and solid states;

4. moisture content (W): referred to in the above definitions is defined as the loss of weight on drying expressed as a percentage of the weight of the dry sample.

Three further indices can be derived from the primary measurements listed above:

1. plasticity index (I_P): the range of moisture content within which a soil is plastic; that is: $W_L - W_P$;

2. liquidity index (I_L): the excess of natural moisture content (W_N) above the plastic limit, expressed as a ratio of the plasticity index; that is:

$$I_L = \frac{W_N - W_P}{I_P} \qquad (3.11)$$

3. activity number (A): introduced by Skempton (1953) and modified by Seed et al. (1964) to be defined as:

$$A = \frac{I_P}{c - n} \qquad (3.12)$$

where I_P is the plasticity index,
  c is the percentage of material finer than
  0.002 mm diameter, and
  n is a number approximately equal to 5 in
  natural soils

One other measure of significance for slope stability is sensitivity (S_t). It was the subject of a comprehensive study by Skempton and Northey (1952) but was first defined by Terzaghi (1944) as:

$$S_t = \frac{q_u}{q_{ur}} \qquad (3.13)$$

where q_u is the unconfined compressive strength
  of the undisturbed soil, and
  q_{ur} is the unconfined compressive strength
  of a remoulded soil at the same water
  content

97

This index measures the loss of strength caused by disturbance and remoulding. However, as extremely sensitive soils become liquid on remoulding, it is difficult to carry out a test for the unconfined compressive strength on the disturbed sample. Consequently, Smalley and Bentley (1980) advocate the use of Soderblom's (1969) measure of sensitivity (S_t):

$$S_t = \frac{H_3}{H_1} \qquad (3.14)$$

where H_1 is the relative strength value determined by the Swedish fall-cone test (Hansbo 1975) on completely remoulded material, and

H_3 is the relative strength value determined by the same test on undisturbed material

Terzaghi and Peck (1967), using the original definition, classified the range of sensitivity as:

most clays	2 to 4
sensitive	4 to 8
extra sensitive	8 to 16
quick clays	>16 (Skempton and Northey 1952)

(values of up to 1500 have been recorded - Penner 1963)

Loss of strength on shearing in coarse soils has been measured by a brittleness index (I_b) (Bishop 1973):

$$I_b = \frac{T_f - T_r}{T_f} \times 100\% \qquad (3.15)$$

where T_f is shear stress at failure, and
T_r is residual resistance to shear

High I_b values represent a large loss of strength within the sheared material and are therefore similar to high sensitivity values in indicating a potential for fluid mass movement. The brittleness index for coal tip material involved in the Aberfan flow slide ranged from 50% to 60% (Bishop 1973).

The concept of 'rapidity' in sensitive soils (Soderblom 1974) has significance for the triggering of landslides as it represents the energy-input required to achieve a remoulded state. A measure of this property would also be useful for determining the susceptibility of sand to liquefaction - a form of sensitive behaviour (see Section 3.3.8 SEISMICITY AND OTHER VIBRATIONS).

A final property important to some types of landslides is dispersibility or slaking under the action of water. Dispersion indices are a measure of how readily particles disaggregate in the presence of water and indicate the potential for fluid mass movement as well as susceptibility to erosion by water. Dispersibility is strongly influenced by particle-size, mineralogy and the chemistry of soil water.

Precise definitions and procedures for determining the properties discussed in this section are available in standard soil testing manuals and texts on geotechnical engineering. As inter-operator variability can be high and as different methods exist, it is important to indicate the procedures followed in determining such properties.

Rock. For most natural slopes, the strength of intact rock (Table 3.6) is less able to reflect inherent strength than the property which has become known as 'rock mass strength'. Rock mass strength is a semi-quantitative measure of inherent strength. It recognises that the weakest part of the slope cannot, at present, be identified or represented by the same techniques as conventionally used for determining intact rock strength.

In rock consisting of uniform properties throughout the entire mass, intact rock strength will in fact equal rock mass strength. Hence, shear strength measured from a sample, in this instance, will yield an appropriate value for slope resistance, when stresses on the potential shear plane are taken into account (equation 3.1). What is more, under such conditions of uniformity, the location and inclination of the potential shear plane can be determined accurately by considering the angle of internal friction and the direction of the principal stress axis controlled by slope form (see Section 3.3.4 INCREASE IN SLOPE HEIGHT).

Most slopes, however, do not display uniform properties throughout their mass; intact strength is usually higher than the strength of structural discontinuities, and orientation of the potential shear plane (and hence normal stresses) cannot be anticipated from intact rock strength and slope form. In addition, irregular structure and variation in other properties present difficulties for the calculation of pore-water pressures. Although these conditions present real difficulties in measurement, they do not invalidate the concept of strength embodied in the Coulomb-Terzaghi equation (3.1).

As already mentioned (see Section 3.3.10 WEATHERING AND ACCUMULATION), progress is being made in the measurement of appropriate strength values for joint surfaces and other structural discontinuities but this begs the question as to whether the appropriate surface is being measured. The technique involved in the assessment of rock mass strength go some way to overcoming this problem. In this approach, a number of readily measured parameters serve as surrogates for the terms within the shear strength equation and are used in combination to provide a relative measure of rock mass strength. They include parameters which indicate:

1. reduction of cohesion
   (a) number of structural discontinuities per unit of rock volume, or their spacing; including joints, fractures, bedding planes, foliations and faults;
   (b) openness or width of discontinuities;
   (c) lateral and vertical extent of discontinuities;
2. cohesion
   (a) strength of intact rock (Table 3.6);
   (b) weathering grade of intact rock (Table 3.3);
3. angle of internal friction
   (a) frictional resistance along discontinuity;
   (b) waviness and roughness along discontinuities;
4. lateral and vertical transient pressures
   (a) movement of water through the rock;
   (b) fillings; that is, contents of joints and voids (also influences apparent cohesion);
5. stress distribution throughout slope
   (a) residual stresses;
   (b) inclinations and orientations of discontinuities with respect to the free face of the slope;
   (c) the stratigraphic position of material with different stress/strain behaviour;
   (d) the stratigraphic position and orientation of material and discontinuities with different permeability.

Of all these factors, the inclination and orientation of structural surfaces have the most clearly demonstrable effect on stability. For example, in a rock slope in which structural surfaces dip outwards towards the free face at an angle less than the slope face, stability is much lower than for similar slopes with inward-dipping structures (**Figure 3.17**). In their natural setting, less stable slopes controlled by these factors develop at lower angles that the more stable slopes. According to Young (1972) these differences are the most common cause of valley-side asymmetry. The difference in stability arises not only because of different amounts of lateral support but also because outward dipping structures tend to direct percolating water to the weaker outer zones of the slope. Terzaghi (1962) has demonstrated that, in slopes with outward dipping strata, the limiting hillslope angle is equal to the angle of dip, in situations where the angle of dip is greater than the angle of friction on the structural surface. However, where joints and discontininuities have no pronounced orientation, the limiting hillslope angle is equal to the angle of internal friction. This angle is controlled by the tightness, interlocking and frictional properties of the joint-blocks and in some rocks may be in excess of 70°.
   Low overburden pressures, weathering and pressure release by erosion result in greater joint space within the outer

Figure 3.17   A rock slide which took place on outward dipping
joint planes; <u>in situ</u> joint surface visible top-
right

Table 3.8   Geomorphic Rock Mass Strength Classification and Ratings (Selby 1980)

PARAMETER	1 Very Strong	2 Strong	3 Moderate	4 Weak	5 Very Weak
Intact rock strength (N-type Schmidt Hammer 'R')	100–60 r:20	60–50 r:18	50–40 r:14	40–35 r:10	35–10 r:5
Weathering	unweathered r:10	slightly weathered r:9	moderately weathered r:7	highly weathered r:5	completely weathered r:3
Spacing of joints	>3 m r:30	3–1 m r:28	1–0.3 m r:21	300–50 mm r:15	<50 mm r:8
Joint orientations	Very favourable. Steep dips into slope, cross joints interlock r:20	Favourable. Moderate dips into slope r:18	Fair. Horizontal dips, or nearly vertical (hard rocks only) r:14	Unfavourable. Moderate dips out of slope r:9	Very unfavourable. Steep dips out of slope r:5

PARAMETER	1 Very Strong	2 Strong	3 Moderate	4 Weak	5 Very Weak
Width of joints	<0.1 mm r:7	0.1-1 mm r:6	1-5 mm r:5	5-20 mm r:4	>20 mm r:2
Continuity of joints	none continuous r:7	few continuous r:6	continuous, no infill r:5	continuous, thin infill r:4	continuous, thick infill r:1
Outflow of groundwater	none r:6	trace r:5	slight <25 $1/\text{min}/10$ m^2 r:4	moderate $25-125$ $1/\text{min}/10$ m^2 r:3	great >125 $1/\text{min}/10$ m^2 r:1
Total rating	100-91	90-71	70-51	50-26	<26

Note: r is the maximum rating obtainable in each class

parts of hillslope compared with the interior of the slope. Consequently the angle of internal friction, as well as normal stress, decrease towards the slope surface. Increased void space also permits a higher water content which can lead to softening of material and a reduction of cohesion in clay-rich rock. Such conditions may facilitate shallow instability on a slope which is stable with respect to deep-seated movements. As noted earlier, the concepts of deep-seated and shallow instability apply also to slopes consisting of cohesive soils.

Increase in void space and consequent lowering of internal friction also results from overburden removal by excavation. In some cases the full reduction of resistance in response to unloading may take several decades (Skempton 1948; Eyles et al. 1978). If completely jointed material has been cut at an inclination close to the angle of internal friction the slope may experience a number of discrete periods of instability, weakening progresses after each period of adjustment. The reduction of internal friction and consequent lowering of the limiting angle for stability will continue until some residual angle of internal friction is attained.

The stratigraphic relationship of different material within the slope also has a marked influence on stability. Its effect on underlying support and the generation of porewater pressures has been discussed in previous sections.

There are a number of methods for combining rock mass strength parameters to represent inherent strength. A two-parameter scheme used to record conditions encountered in an investigation tunnel through greywacke is shown in **Figure 3.18**. Six arbitrary classes have been derived on the basis of weathering grade (Table 3.3) and the intensity of defects which have resulted from jointing and faulting; class 0 represents the weakest conditions and class 5 the strongest.

Selby (1980) has developed a more comprehensive system which he has used in a number of different environments to determine the influence of rock mass strength on land form characteristics (Selby 1982a; Selby 1982b; Moon and Selby 1983). It includes eight of the 13 rock mass parameters listed previously to produce seven parameter categories. Each one is rated on a five point scale and the rock mass strength is indicated by the sum of the ratings **(Table 3.8)**.

Dilatancy and peak strength.   Dilatancy is the process of volume increase experienced by certain materials when they are disturbed and sheared. Volume increase may also result simply from unloading. In clays, this process is enhanced by pressures resulting from the intake of water. Densely-packed frictional material, jointed rock, silts and overconsolidated clays all exhibit some degree of dilatancy.

During shear testing dilatancy is recorded as axial strain in a direction opposed to the normal load. Prior to maximum expansion the material initially offers peak

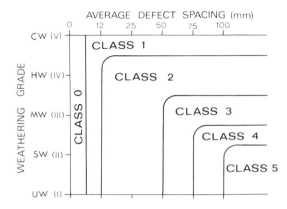

Figure 3.18   Rock mass strength classification by weathering and intensity of defects (Riddolls and Perrin 1975). For definition of weathering grades see Table 3.3

resistance to shear stress (θ'_m) followed by decline to a residual resistance (θ'_r) as dilatancy occurs. The most common explanation for this process emphasises the necessity for tightly packed grains to separate and ride over opposing grains before shearing can take place between them. When grains on one side of the shear surface have 'cleared' those on the other side, maximum expansion has been achieved and resistance to shearing is reduced. Orientation of some grains parallel to the shear stress may also further reduce resistance before θ_r is achieved.

From this explanation it is evident that roughness of the shear surface, particle shape and density of packing will determine the amount of dilatancy that takes place. These factors together with the magnitude of normal load and frictional properties of the surface are responsible for generating peak resistance. Residual resistance, however, depends on how the particles readjust to normal load after shearing.

The readjusted state differs most radically from the presheared state in overconsolidated clays. An overconsolidated material (as opposed to normally consolidated) is one which at some time in the past has been subject to greater loads than those experienced at present. Former loading results from

105

consolidation pressures during deposition and burial, cryostatic pressures and wetting and drying stress. When these are diminished by erosion or changes in climate, an overconsolidated state is apparent in the affected material.

Some materials including fissured overconsolidated clays, loosely-packed sediments, saturated hillslope soils, fault gouge and other previously sheared material already exist at or near their residual strength. In this state any true cohesion that may have originally existed has generally been completely eliminated and when the material is saturated the influence of apparent cohesion is also negligible. Thus in these materials shear strength (s) can be represented by the expression:

$$s = \sigma' \tan \Phi'_r \qquad (3.16)$$

where σ' is effective normal stress, and
Φ'_r is the residual angle of internal friction with respect to effective normal stress

Although shearing may not always be necessary for reducing all types of material to a residual state (Skempton 1970), some degree of expansion is always associated with the reduction of peak strength. This mechanism involves an increase in void ratio and a consequent reduction in particle contact. As a result the frictional, electrical and molecular forces offering resistance are diminished. Initially increased void space may increase resistance by inducing high negative pore pressures but these will eventually be destroyed as water permeates the material.

A comprehensive treatment of this topic is provided by Skempton (1964). Other references are given in Section 3.3.10 WEATHERING AND ACCUMULATION.

Influence on landslide form. Most of the discussion in this section has dealt with the influence of material properties and structure on inherent strength and slope stability. However the form and type of landslide produced is also influenced by the characteristics of the constituent material as indicated by the commonly occurring relationships illustrated in **Figure 3.19.**

Indeed, nearly all the factors which influence rock mass strength also exert some control over landslide morphology. However, in rock, as opposed to soil, the presence, distribution and orientation of structural discontinuities usually exert an overriding influence over the other factors. In massive or randomly jointed rock the depth and location of weathering is important and closely related to the depth and volume of movement. In rocks with pronounced structure, on the other hand, the depth, volume and coherence of the displaced mass relates less to weathering and more to the relationship between slope geometry and joint patterns.

In cohesive soils the propensity for shallow, translational flow is dependent upon the strength of interparticle bonding and the ability of the constituent material to take up water. In the absence of strong structural control, the type and volume of movement depends to a large extent on whether the development of relief has been sufficient to allow critical height to be achieved for the unweathered material (see Section 4.2 FLUVIAL DOWNCUTTING). Landslides occurring on a slope of less than critical height have their volume controlled by depth of weathering and the nature of their translational movement controlled by the uptake of water and slope angle. The form of the land surface and the vegetation cover also exert some influence on the morphology of the displaced mass under these conditions. In the case of slopes at critical height, the shape and position of the failure plane develop in response to the distribution of principal stresses which are in turn influenced by the geometry of the slope.

Figure 3.19   The influence of material and structure on landslide form (key to diagrams on following three pages)

Ia		Ib	
1	rock slump	1	block glide
2 and 3	rock slides	2	wedge failure (rock slide)
4	slab failure	3	rock slide
5	granular disintegration	4	rock topple

II			
1	earthflow (discrete)	4	quick slide (flow slide)
2	mudflow (fast earthflow)	5a and b	earth slump
3	extensive earthflow	6	earth block slide

III			
1	debris block slide	4	dry ravel
2a and b	debris avalanche (slide)	5	debris slump
3	debris flow (torrent)		

I

ROCK

a. Massive or Random Structure

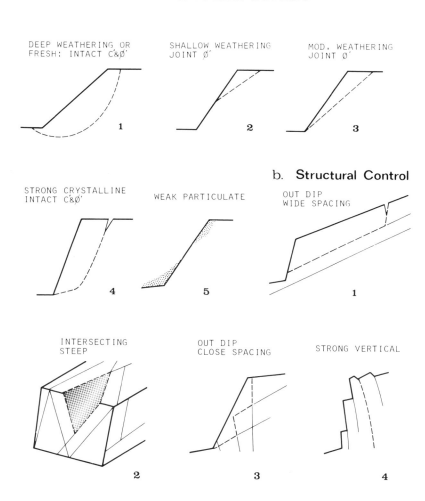

DEEP WEATHERING OR
FRESH: INTACT C&Ø'

1

SHALLOW WEATHERING
JOINT Ø'

2

MOD. WEATHERING
JOINT Ø'

3

b. Structural Control

STRONG CRYSTALLINE
INTACT C&Ø'

4

WEAK PARTICULATE

5

OUT DIP
WIDE SPACING

1

INTERSECTING
STEEP

2

OUT DIP
CLOSE SPACING

3

STRONG VERTICAL

4

II

COHESIVE SOILS

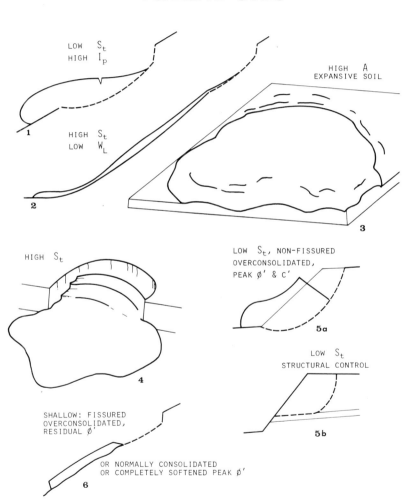

LOW S_t
HIGH I_p

1

HIGH S_t
LOW W_L

2

HIGH A
EXPANSIVE SOIL

3

HIGH S_t

4

LOW S_t, NON-FISSURED
OVERCONSOLIDATED,
PEAK ϕ' & C'

5a

LOW S_t
STRUCTURAL CONTROL

5b

SHALLOW: FISSURED
OVERCONSOLIDATED,
RESIDUAL ϕ'

OR NORMALLY CONSOLIDATED
OR COMPLETELY SOFTENED PEAK ϕ'

6

109

III

LESS COHESIVE SOILS

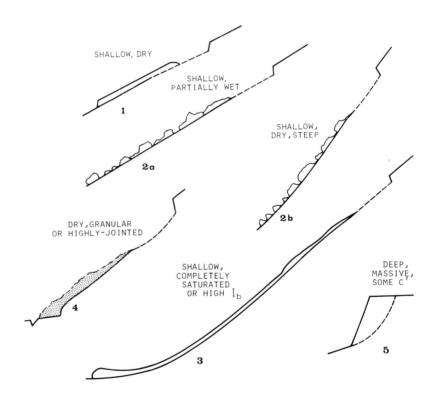

S_t	-	SENSITIVITY	A	-	ACTIVITY
I_p	-	PLASTICITY INDEX	I_b	-	BRITTLENESS INDEX
W_L	-	LIQUID LIMIT	ϕ'	-	INTERNAL FRICTION
			C'	-	COHESION

ENVIRONMENTAL AND GEOMORPHIC MODELS FOR THE DEVELOPMENT OF UNSTABLE TERRAIN

4

Why does a particular terrain or slope become unstable? To answer this question it is necessary to look beyond the geomechanical conditions that exist at the time of failure to the history of the slope and its environment. Whereas Chapter Three identified specific destabilising factors and failure mechanisms, this chapter examines the evolution of conditions which allow those factors and mechanisms to be effective; that is, the development of instability.

In the human setting, the development of instability is easy to appreciate. Mechanical aids permit dramatic changes to be made to the slope. Excavation, which is the instrument of urban development, can shift a hillslope with stable height/slope relationships towards its critical threshold within the space of a few hours. Modern technology can rapidly concentrate large weights and quantities of water onto the ground and can translocate or radically alter surficial material. These and other modifications to the slope can produce instantaneous failure or induce a condition of instability that will ultimately lead to failure (Eyles et al. 1978).

The development of instability in the natural environment, on the other hand, is a much more complex issue. Several models are proposed to explain this phenomenon and they involve respectively the following geomorphic and environmental processes: tectonic uplift; fluvial downcutting; weathering-related changes in material properties and depth of regolith; climatic change; vegetation change; and previous geomorphic activity. The various models emphasise the critical role of these processes in the development of a state of marginal stability in which landslides can be readily triggered by existing transient factors. The explanations offered are not necessarily mutually exclusive; they are instead appropriate to different scales of time and space.

4.1   TECTONIC UPLIFT AND REGIONAL INSTABILITY

Over long periods of geologic time, the development of instability can be viewed as a function of relief[1] produced by
major tectonic uplift. Although relief also varies with
eustatic (sea-level) changes, epeirogenic movement[2] and
isostatic adjustments[3], these factors produce little change
compared to that brought about by major tectonic uplift. For
example, the maximum glacial/interglacial sea-level oscilla-
tion in the last 20 000 years has been only 130 metres
although 80 million years ago sea-level is thought to have
been 500 metres above its present level (Fairbridge 1978).
These changes represent less than one-twelfth of the altitude
attained by high alpine regions which have been subject to
major tectonic uplift. It is true, however, that a rise in
sea-level may initiate coastal instability and that a fall in
sea-level may destabilise inland slopes by way of river rejuvenation, but these mechanisms are most effective where there
is already appreciable relief.

Uplift of mass to high altitudes increases the potential
gravitational energy available to drive processes, slopes are
steepened, weight is directed increasingly into shear stress,
and material in motion releases high amounts of kinetic
energy.

Major tectonic uplift also subjects material to the
severe atmospheric conditions associated with high altitude
and creates local climatic patterns of its own (**Figure 4.1**).
Orographic precipitation, low evapotranspiration and
suppressed vegetation growth which occur at high altitude
ensure high runoff occurs and fluvial processes are active.
Fluvial action in turn is responsible for developing local
relief, for undercutting slopes and for removing slope deposits which might otherwise impede mass movement.

The importance of the slope and height elements of high
relief is obvious from the discussion in Sections 3.3.4
INCREASE IN SLOPE HEIGHT and 3.3.5 LATERAL SUPPORT AND
SLOPE ANGLE. However, the height element of relief is also
critical to the maintenance of instability even when friction
is the only operative strength parameter. If the removal of mass

---

[1] Relief is defined here as the difference in altitude between
sea-level and the highest point within a unit area of land.
Thus an area of high relief is characterised by higher altitude and steeper average slope than an area of low relief.
[2] Broad vertical movement of the earth's surface without
folding or crumpling of strata.
[3] These are movements which compensate for anomalies in the
height/density balance among adjacent parts of the earth's
crust.

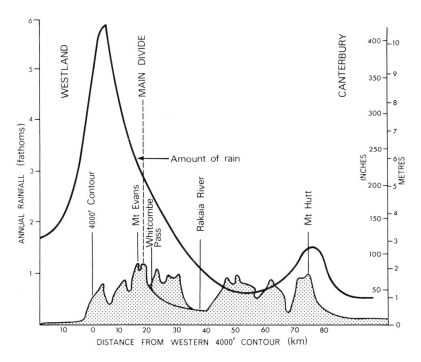

Figure 4.1   Concentration of rainfall in a section across
the Southern Alps, New Zealand (Chinn 1979)

movement deposits is restricted, stable debris-slopes extend
upslope thereby reducing the area available for continued mass
movement.   Thus, the higher the slopes, the longer it takes
for them to be mantled by protective debris and therefore the
more prolonged is the period of instability, compared to areas
of low relief.

      Areas of major tectonic uplift are also associated with
seismicity and, in places, volcanic activity, both of which
can dramatically increase landslide activity.   Indeed over 75%
of the energy released by the world's earthquakes between 1904
and 1952 was concentrated in the vicinity of the mountain
ranges which make up the convergent crustal-plate boundaries
of the Circum-Pacific Zone (Stacey 1969).

4.1.1   CYCLICITY OF INSTABILITY

Uplift of a region does not last forever, as the geologic record attests. Relief, however, continues to be reduced after uplift has stopped; this occurs rapidly at first as landsliding continues and then more gradually as slower forms of mass movement and other erosional processes assume dominance. Thus within this explanatory model, regional instability is presented as a cyclic, relief-dependent, self-annihilating state which, given time and tectonic stability, destroys the conditions necessary for its continuance. In its most extreme expression all parts of the topography can be subject to both shallow and deep-seated landslides. For this degree of instability to exist there must be substantial relief achieved over a wide area. Such conditions pertain in active or recently active orogenic (mountain building) belts of the earth's crust.

The history of orogeny indicates that both episodic uplift and sufficiently long periods of intervening quiescence exist for this scale of instability also to respond cyclically. As Schumm (1963) states '...rapid uplift (that is, uplift rates greater than denudation rates) probably is the rule in orogenic areas, as the existence of mountains attests'. In addition, Fairbridge (1978) notes: 'There have been many long, relatively quiet times during our protracted global history, interspersed with crescendos of activity, marked by volcanicity, crumpling and mountain building...this periodicity (between orogenies) is approximately 200 million years.'

Denudation following the cessation of orogenic activity ultimately reduces relief (albeit unevenly for much of the time) and thereby lowers both applied and internal stress. At some time in this reduction of available energy the stress levels drop below those required to initiate widespread landsliding and the cycle of regional instability is terminated. Schumm (1963) concludes from his analysis of sediment yields and uplift rates that a 200 million year period of quiescence is easily long enough to subdue relief to a peneplain form. In fact he estimates that a peneplain could be developed from a 5000 metre high mountain range in 15 to 110 million years. Long before the gentle peneplain form has been achieved the major landslide cycle (**Table 4.1**) and even localised forms of slope instability will have ceased.

Shorter duration periods of instability such as epicycles, phases, episodes and events (Table 4.1; **Figure 4.2**) may follow or be superimposed on the major cycle to extend or intensify the period of landslide activity. The epicycles caused by isostatic adjustment to mountain erosion, for example, are produced by episodic uplifts which, although individually smaller than the original orogenic uplift, are considered to be sufficient to extend the time to peneplanation five times

Table 4.1   Hierarchy of Landslide Activity

Category	Intervening Period (approximate)	Typical Causes
Cycle	millions of years	orogeny
Epicycle	centuries - millennia	isostatic adjustment eustatic change change in seismic activity climatic change vegetation change
Phase	decades - centuries	variation in intensity of above deforestation
Episode	years - decades	variation in climatic pattern earthquake swarm
Event (multiple occurrence)	week - days	fluctuation in climatic parameter earthquake
Occurrence	minutes - hours	fluctuation in slope hydrology

beyond that required if no isostatic adjustment is taken into account (Schumm 1963).

Continued crustal obduction or subduction caused by forces originating outside the orogenic zone, however, may be more important in maintaining uplift than is local isostacy. In the Southern Alps of New Zealand, for example, continual obduction of Pacific-plate sediments, has allowed, by various estimates, a mass of rock equivalent to a height of 20 to 50 km to be thrust upwards and largely eroded over the last 2.5 million years (Wellman 1975; Adams 1978), even though the mountain range may never have exceeded its present maximum altitude of 3 to 4 km.

4.1.2   RIVER EROSION AND LOCAL RELIEF

The source of load in rivers flowing from tectonically active mountains highlights the importance of fluvial action in developing sufficient local relief to allow mass movement to occur. For example, the Shotover River flowing from the schist zone of New Zealand's Southern Alps carries a total

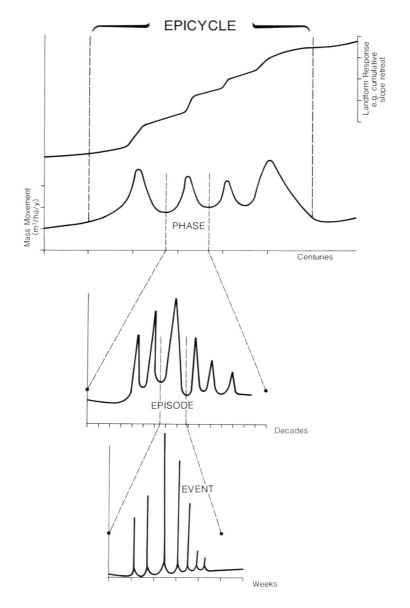

Figure 4.2   Hierarchy of landslide activity and landform response

load estimated at between 400 and 900 $m^3/km^2/y$ (O'Loughlin and Pearce 1982). Analysis of the material involved indicates that as much material comes from slope erosion as it does from river bank and bed erosion: 40% slope erosion, 20% bank erosion, 20% bed degradation, and 10% from a number of minor sources (Ministry of Works and Development 1977). However, because river action operates on a much smaller surface area than do slope processes (particularly in the confines of a mountain range), stream incision will be faster than slope removal - leading, of course, to the formation of local relief.

Potassium-Argon and Rubidium-Strontium rock dating of the schist of the Southern Alps (Sheppard et al. 1975) and analysis of sediment yields (Adams 1978) indicates, that over the last 2.5 million years in this part of New Zealand, mean uplift rates (0.1 to 0.3 mm/y) exceed denudation rates (0.07 mm/y). Under these conditions, if the Shotover River data are representative of the region, it can be concluded that local relief is actively increasing. Schumm (1963) has argued that the disparity between slope lowering and river incision will continue long after uplift has subsided to a rate that is more in balance with the gross denudation rate because of the relief already achieved. A balance between uplift and gross denudation would thus be associated with continued relief development and not with time-independent relief forms such as those postulated by some earlier geomorphologists (Penck 1953; Wegman 1957). Such a balance is thought to have existed for most of the Southern Alps in the last 2.5 million years on the basis of Adams' (1978) calculations which show:

tectonic uplift   = 620 $\pm$ 20 Mt/y
total river load   = 700 $\pm$ 200 Mt/y
offshore deposition = 580 $\pm$ 110 Mt/y
(where Mt = 10^9 kg)

Local relief is developed and maintained in such cases because material is removed unevenly from the landmass. Erosion preferentially and repeatedly occurs in linear zones of the river valleys and often in sub-linear zones on the interfluves.

The stimulus that relief gives to erosion is shown by the relationship derived from United States data by Schumm (1963):

$$Log_{10}D = 26.86H - 2.236 \qquad \textbf{(4.1)}$$

where D is denudation rate in metres per 1000 years and H is relief/length ratio (relief of basin divided by basin length)

The equation indicates that denudation rates increase approximately ten-fold with an increase in the relief/length

117

ratio from 0.005 to 0.05. This is the degree of increase which would be expected as a result of major tectonic uplift of a low-relief area. Some data subsequently obtained from different mountain regions of the world, however, indicate denudation rates higher than those predicted by this equation. These differences are likely to occur as a result of variations in climate, rock mass strength, vegetation cover and other regional factors.

4.1.3 EROSION - THE NEW ZEALAND CASE

The most active orogenic belts in the world are found at convergent or compressive crustal plate boundaries along which continental-margin sediments are thrust and folded to produce high mountain ranges. The Circum-Pacific Zone represents the greatest concentration of present-day orogenic activity and this in turn is associated with extremely active landsliding and erosion. New Zealand, situated on the boundary of the Australian and Pacific crustal plates, illustrates the typically high erosion rates experienced in these areas.

The New Zealand Land Resource Inventory (Eyles 1983) has provided a comprehensive assessment of the areal distribution and severity of the types of erosion recognised by Campbell (1951). The inventory recorded the presence and condition of active or exposed erosional forms within nearly 90 000 individual map units having a median area of 154 hectares. **Table 4.2** lists the areal extent of mass movement types and **Figure 4.3** provides an example of mapped distribution for debris avalanches which constitute one of the landslide types included in the survey. The same study indicates that approximately 52% of the land area covered by mapping units also exhibits surface erosion (sheet, wind and scree).

As the New Zealand Land Resource Inventory does not record the age or volume of erosion forms (apart from designating them as 'present' features) it is impossible to arrive at a displacement rate for mass movement from the available data. There is also little information on how much of the material displaced by mass movement enters the drainage system at the time movement occurs. There is some indication, however (Crozier et al. 1980; O'Loughlin and Pearce 1982), that contemporary mass movement events have a low sediment delivery ratio; that is, only a small part of the material displaced is immediately transported out of the catchment. The displacement rates for some New Zealand landslide episodes are compared with those from other areas in **Table 4.3.**

The provision and availability of load to New Zealand rivers vary greatly from place to place and denudation rates derived from down-basin sediment yields may bear little relationship to slope erosion higher in the basin. In some areas mass movement deposits are being stored in headwater tributary valleys (Crozier 1983), in other parts of the country rivers

Table 4.2 Area of Map Units Affected by Mass Movement Erosion Types in New Zealand (Eyles 1983)

Erosion Type	North Island		South Island		New Zealand	
	Area (ha)	%	Area (ha)	%	Area (ha)	%
Soil slip	3 397 000	30.0	3 615 800	24.0	7 012 800	26.0
Earth slip	280 300	2.5	58 500	0.4	338 800	1.0
Debris avalanche	1 218 900	10.7	1 603 000	10.6	2 821 900	11.0
Earthflow	1 011 500	8.9	33 300	0.2	1 044 800	4.0
Slump	65 800	0.6	44 100	0.3	109 900	–
Total mass movement*	5 038 200	44.1	4 602 000	30.5	9 640 200	36.0

* The total area affected by the five erosion types is different from the sum of the individual types because up to three types can be recorded in a mapping unit.

0 _____ 200
kilometres

Figure 4.3   The distribution of debris avalanche erosion in
New Zealand (Eyles and Eyles 1982)

are obtaining load previously stored by Pleistocene fluvio/
glacial processes (Mosley 1978) while most rivers appear to be
starved of load until a mass movement event occurs (Adams
1979). These factors suggest that the prediction of river
load by single factors, such as mean annual precipitation
(McSaveney 1978) or basin slope (equation 4.1), can provide
only an approximation of actual yields.

As river measurement records in New Zealand have
improved, a more accurate picture of erosion in a tectonically
active area is being obtained **(Table 4.4)**. The New Zealand
river data presented in Table 4.4 are derived from about 20

Table 4.3a   Volume of Material Displaced and Area Eroded During Periods of Landslide Activity
(Crozier et al. 1982)

Volume (m³/ha)	Area surveyed (ha)	Area eroded (%)	Period of episode	Locality	Source
1150	24 000	25	1970	Adelbert Ra, Papua NG	Pain and Bowler 1973
844	143	18.2	1968-69	San Dimas, California (pasture)	Rice and Foggin 1971
100-800	2000-50	-	1966	Mangawhara Valley, NZ	Selby 1976
690	23	9.7	1977	Pakaraka, Wairarapa, NZ	Crozier et al. 1982
506	216	1.5	1980	Wainitubatolu, Fiji	Crozier et al. 1981
400	280	-	1973	Matahuru and Mangapiko, NZ	Selby 1976
337	112	5-12	1966-67	Bell Canyon, California	Rice, Corbett and Bailey 1969
298	145	5.8	1968-69	San Dimas, California (scrub)	Rice and Foggin 1971
125	550	<10	mid 1960's	Notown, West Coast, NZ	O'Loughlin and Pearce 1976
77	162	1	1971	Hawkes Bay, NZ	Eyles 1971
26	1 267	0.3	1976	Stokes Valley, NZ	McConchie 1977
6.3	5 020	-	1974	Ashland Creek, Oregon	Smith and Hicks 1982

Table 4.3b Long-term Displacement Rates for Debris Slides and Debris Avalanches

Rate (m³/km²/y)	Period of Record (y)	Area surveyed (km²)	Locality	Source
Dominantly Forest				
1500	25	720	Redwood Ck, California	Kelsey et al. 1981
280	35	575	Van Duzen R, California	Kelsey 1980
100	6	5.5	North Westland, New Zealand	O'Loughlin and Pearce 1976
71.8	84	19.3	Olympic Pen, Washington	Fiksdal 1974
47	23	50.2	Ashland Ck, Oregon	Montgomery in Smith and Hicks 1982
45.3	25	12.3	Aldar Ck, Western Cas, Oregon	Morrison 1975
37.2	25	21.4	Andrews Fst, Western Cas, Oregon	Swanson and Dyrness 1975
11.2	32	246	Coast Mtns, British Columbia	O'Loughlin 1972
8.8	10	6	Big Beef Ck, Washington	Madej 1982
Disturbed: Clearcut/Scrub/Pasture				
1000-4000	3	6.2	North Westland, New Zealand	O'Loughlin and Pearce 1976
1500-3000	**	1.6	Hawkes Bay, New Zealand	Eyles 1971
2850	48	1350	Wairarapa, New Zealand	Crozier 1983
1000	**	20	Mangawhara, New Zealand	Selby 1976
350*	22	1.7	Lone Tree Ck, California	Lehre 1982
161	25	7.9	Andrews Fst, Western Cas, Oregon	Swanson and Dyrness 1975
125*	100	1.7	Lone Tree Ck, California	Lehre 1982
117	15	4.5	Aldar Ck, Western Cas, Oregon	Morrison 1975
29	10	17	Big Beef Ck, Washington	Madej 1982
25	32	26.4	Coast Mtns, British Columbia	O'Loughlin 1972
25	**	12.7	Stokes Valley, New Zealand	McConchie 1977

* Weight to volume conversion using 1.5 g/cm³   ** Calculated on return period of triggering storm

Table 4.4   Sediment Yield ($m^3/km^2/y$) from Steep Terrain

(a) New Zealand Rivers (Thompson and Adams 1979; Adams 1979)[1]

	Interquartile Range	Median	Range	Sample Size
South Island[2]	75-914	294	25-4861	26
North Island	128-833	267	19-15 556	44

(b) Characteristic Rates for New Zealand Mountains
(O'Loughlin and Pearce 1982)

Dry climate low schist	<20
Marlborough and North Island greywacke	1000-2000
Glaciated North Otago, Canterbury and Marlborough	50-5000
Wet climate schist Alps	1000-10 000

(c) World Rates (Young 1974)

	Interquartile Range
Steep relief	92-970

---

[1] The data provided by these authors have been converted from tonnes to cubic metres by a bulk density of 1.8 tonnes/m^3

[2] Suspended sediment only. However, in general, suspended sediment makes up 93% of total load of South Island rivers draining to the ocean, dissolved load 4%, bedload 3%. The proportions for North Island rivers are 88%, 10% and 2% respectively (Adams 1979).

years of record. By drawing on other lines of evidence, Adams (1979) argues that the rates shown for South Island rivers are only about half those of the long-term rate for the region. According to to Adams, low frequency, earthquake-triggered, landslide episodes substantially increase suspended sediment loads over the long term. Climatically-triggered landslide episodes may be an even more important source of pulse sediment supply.

Adams' (1979) observation that measured river sediment yields represent only half the long-term, geological rate is important in respect of the widely held view that New Zealand is currently experiencing unusually high, accelerated rates of

Development of unstable terrain

erosion as a result of deforestation, land use practices and increased storminess (Grant 1983). Indeed **Table 4.5** suggests that landslide activity for instance, has been increasing, at least in some parts of the country.

Table 4.5 Landslide Displacement Rates ($m^3/km^2/y$) for Pakaraka Experimental Catchment, Wairarapa, New Zealand (Crozier 1983)[1]

Period	Site 1	Site 2	Site 3	Mean
AD 1961-AD 1977	11 900	10 100	2630	8210
AD 1932-AD 1979		Regional Rate		2850
533 BP-AD 1961	240	170	30	150
1580 BP-533 BP	230	130	40	130
3290 BP-1580 BP	30	–	–	30

[1] All data derived from field measurement and drilling in a 23 ha hill country catchment, except for the regional data which were obtained by air photo analysis of a 1350 km^2 area of pasture-covered hill country.

If the generally accepted conditions of supply-limited river load, and episodic slope supply are operating and yet measured, short-term yields are only half those of long-term yields, then **Figure 4.4a** would represent a basic model for erosion of much of New Zealand's southern mountainland. The model indicates that current rates result from measurement during a period of 'chronic' as opposed to 'pulse' sediment transport.

If, however, it is also accepted that slope erosion and landslide activity have increased in the last few decades, then three possible explanations present themselves:
1. chronic sediment transport has indeed increased in response to recent slope instability but has only recently reached a rate equivalent to half the long-term rate (**Figure 4.4b**). It follows that chronic sedimentation is 'normally' less than half the long-term rate and that pulse sediment supply is the major source of river sediment;
2. sediment from recent, accelerated landslide activity is being stored in upper catchments and has not yet been incorporated or measured in the main river;
3. pulse and chronic sources of sediment have not changed their relative contributions but have deccelerated in concert (**Figure 4.4c**). In such a case, increased slope activity over the last few decades produces a rate of chronic sedimentation which is

124

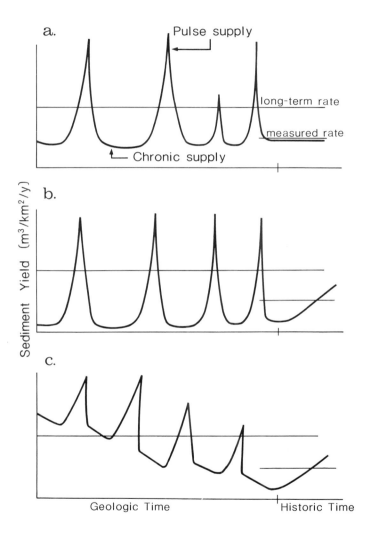

Figure 4.4   Possible explanations for changes in erosional activity in New Zealand

less than earlier chronic rates. This model may be appropriate to an area where tectonic uplift has ceased and which is experiencing a reduction in relief. However it is unlikely to apply to an area such as New Zealand which is undergoing active tectonism.

For the reasons given in this section the high rates of mass movement and erosion in New Zealand can be related directly or indirectly to active and recent tectonic uplift. The presence of high mountain ranges extending across the path of moisture-laden, mid-latitude, westerly winds ensures that heavy orographic rainfall and effective fluvial processes occur. In addition, both the greywacke and schist rocks which make up the axial ranges are often intensely folded, jointed, foliated and faulted. The remaining rock types are for the most part relatively weak, Tertiary sediments which are unable to sustain high relief in the relatively wet conditions which prevail. Alternating sequences of permeable and impermeable strata aggravate the inherent instability of these sediments and tilted strata often present unfavourable dips within the highly dissected terrain.

Two other forms of major disturbance which will be discussed in more detail later, also contribute to the inherent instability of New Zealand's terrain. The first is the wide extent of airfall deposits, both loess and volcanic tephra, which have mantled the slopes in recent geologic time. As these are relatively permeable deposits compared to their depositional surfaces, they readily develop perched watertables and become highly unstable. The second major destabilising influence has been the large-scale deforestation which accompanied Polynesian settlement of New Zealand and which was dramatically accelerated as a result of European land use practices in the last 150 years.

The inherent instability produced by these factors provides a sensitive terrain within which extreme weather conditions and seismic activity are particularly effective in producing the high rates of landslide activity and erosion that are experienced in New Zealand.

4.2   FLUVIAL DOWNCUTTING

Figure 4.5 represents an interpretation of an instability model that was first proposed by Skempton (1953) and subsequently discussed by Carson and Kirkby (1972). It suggests that four quasi-equilibrium slope forms may develop in response to the different mass movement processes which assume dominance at certain stages during and after stream incision.

In stage 1, a 'V'-shaped valley is formed by stream incision and shallow sliding. Sliding takes place on a planar

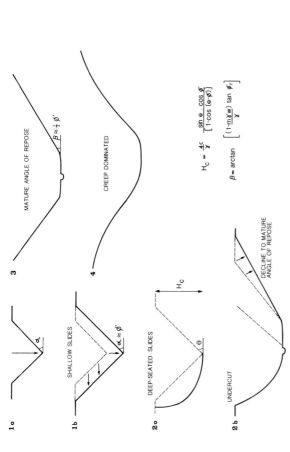

Figure 4.5 **Fluvial downcutting and valley-side instability** (after Skempton 1953);
α, β and Θ represent critical slope angles at different stages of the model,
γ_w is the unit weight of water, $\emptyset'r$ is the residual angle of internal friction,
m is the height of the water table above the slip surface and other symbols are
those used in equation 3.1

surface at a depth dictated by the degree of weathering and water-induced softening of the slope material and, as a consequence, parallel retreat of the slope occurs. As weathering-related processes reduce true cohesion to zero, the sole strength parameter is the angle of internal friction. Active stream incision at the toe of the slope triggers mass movement and thereby maintains slope angles close to the angle of internal friction (**Figure 4.5: 1b**). This process will continue until valley incision reaches the critical depth (H_c) for the onset of deep-seated instability (**Figure 4.5: 2a**). The progressive removal of lateral support by stream incision allows increasingly greater stress to be imposed at a depth and, when this exceeds shear strength, deep-seated failure representing stage 2 occurs. Landslide debris forces the channel to undercut the opposite valley side and further deep-seated sliding and valley widening takes place.

Valley widening by this process and cessation of stream incision allows slopes to develop what Skempton refers to as a 'mature angle of repose'. This will appear unevenly throughout the valley, occurring initially at sites on the inside of river bends and at other places free from stream undercutting (**Figure 4.5: 2b**).

The mature angle of repose signifies stage 3 of the model (**Figure 4.5: 3**) and is achieved by shallow sliding triggered by porewater pressures at saturation. Skempton and DeLory (1957) and others have subsequently shown that with low cohesion or if only residual strength is mobilised and groundwater is assumed to reach the surface, then the limiting angle for stability will be approximately half the residual angle of internal friction.

The mature angle of repose can be expected to occupy a relatively long time span in this model as processes leading to its development are relatively rapid and the creep or wash processes which subsequently modify its form (**Figure 4.5: 4**) are comparatively slow. Given a wide range of topography, commonly-encountered slope angles may represent the critical angles for shallow slides of stage 1, deep-seated slides of stage 2, and the mature angle of repose (**Figure 4.6**). Because of its persistence however, the mature angle of repose is likely to be encountered as the dominant or characteristic hillslope angle.

4.3   IN-SITU CHANGES OF PROPERTIES

Hong Kong is underlain by a mantle of residual granitoid that is about 30 metres thick. As Durgin (1977) notes, this material represents the most advanced stage of granitoid weathering and is commonly found in the tropics where leaching of rock is more important than is surface erosion. In 1966, a

Figure 4.6   Deeply incised terrain representing surficial
              movements of stage 1 (background) and the
              deeper-seated movements of stage 2 (fore-
              ground), Weber County, New Zealand

rainstorm affecting this material triggered the most disastrous mass movement ever recorded in the area (So 1971). If the accumulation of weathering products can only occur in the absence of effective erosion, why did the conditions change to permit a sudden and large mass movement event? It is tempting to answer this question by invoking that, because of the high magnitude of the triggering rainfall event, a shift in weather patterns was the cause, implying that rainfalls of such high magnitude had never been attained before the 1966 event. However, although the intense rainfall was undoubtedly the triggering factor, it is highly unlikely that similar or larger rainstorms had not occurred on a number of occasions in the geological past. In the absence of tec- tonic uplift, the most likely process responsible for the development of instability in this case is the progressive maturing or 'ripening' of the slope to a stage where external factors could be effective in triggering movement. In this case, slope-ripening occurred by the weathering-related changes described in Section 3.3.10 WEATHERING AND ACCUMULATION. Models proposing such slope-ripening have been employed to explain the development of instability in both hard rock and regolith.

4.3.1 ROCK SLOPE CHANGES

The development of instability on steep slopes consisting of hard, jointed-rock has been observed by Eyles et al. (1978) in their study of landslides in Wellington City, New Zealand. The city has been progressively established over the last 100 years, mostly on steep greywacke slopes. Streets at a density of up to 160 metres/hectare and associated building platforms have been created by cut-and-fill methods. Most slopes have been cut from the bedrock at an angle of 70°, conforming to the steepest batter allowed under City Council by-laws. City records and other historical sources permitted broad dating of suburban development and major landslide events.

Slope movements on rock faces within the city are predo- minantly shallow, low-volume, highly-disrupted rock slides. In any one mass movement event, the location appears to be related to slope angle, the nature of jointing (partly related to aspect) and in places to the presence of localised slope drainage. The lack of correlation between instability and slope height and the presence of weathering features on the slide surface indicate that intact rock cohesion is not an operative strength factor.

There are four recognisable stages in the development of these slopes in Wellington City: a relatively stable, slope- ripening stage; a rock-slide adjustment stage; a regolith development stage; and a debris slide adjustment stage. The first stage commences at the end of construction although natural analogues also exist in the region. Slopes are cut at an inclination of 70°, generally more than doubling their pre-excavation angle. At the time of excavation, the water-table will still be adjusted to the original slope form but will gradually subside as drainage occurs from the cut. Despite the new angle and high water-table, slopes are generally 'stable' at the end of construction. If they show signs of instability, they are cut back further at the same angle until material of sufficient strength is encountered. Where continued excavation cannot be carried out artificial support is employed.

In many cases the cuts will have been formed in slightly or moderately weathered greywacke (grades II and III, Table 3.3). Martin and Millar (1974) have demonstrated that the joint-strength of this material yields angles of shearing resistance (internal friction) between 45° and 70°, at normal loads typical of Wellington rock slides (<250 kPa). Under these relatively low normal loads, joint blocks are not broken during failure; instead tightness and geometry of interlocking blocks provide most of the frictional resistance. The tightly-keyed joints and high joint friction of freshly cut slopes offer sufficient resistance to support angles of 70°.

Apparently no major landslide events occurred within the first 50 years of Wellington's development (Eyles et al. 1978). Although this can be partly attributed to the young age of slopes, other factors include the low incidence of critical rainfall conditions and the relatively low density of cut slopes during the early stages of the city's development. Recent observations indicate that the 'stable phase' can be much shorter than 50 years.

As a slope ages, active joint processes operate within the cut face and progressively open up the joint-system. This effect may be enhanced by dilation of the rock mass in response to the initial removal of overburden. The net effect of these processes is an increase in void space within the surface zone of the rock mass and consequently a lowering of the angle of shearing resistance. Increase in void space also provides an opportunity for cleft-water pressures to build up during an intense rainstorm. Aging of the slope in this manner can be viewed as slope-ripening, where resistance is reduced to a level which can be periodically overcome by applied stress of the prevailing climatic regime.

The onset of episodic landsliding characterises the start of the second stage of slope development. On average, slopes younger than 40 years produce rock slides approximately three times the volume of those from older slopes. Some older slopes, although scarred by previous rock slides, remain 'stable' even under high magnitude, triggering events. Thus it can be implied that slopes in the rock slide adjustment stage initially experience relatively large-volume slides and subsequently less frequent lower-volume slides and that they lower their slope angle in response to decreasing angles of shearing resistance. For example, most of the rock slides that occurred in Wellington during the wet winter of 1974 produced slopes ranging between 47° and 63°. These post-sliding angles correspond approximately to the 45° to 55° angles of shearing resistance for joint surfaces of weathering grade IV.

The volume of rock slides also appears to increase with the extent of the vegetation cover. Thus root cohesion may be important in allowing slopes to be temporarily supported at angles greater than the angle of shearing resistance.

The lowest stable slope angle, with respect to rock slides in highly weathered (grade IV) greywacke, appears to correspond to a 'residual' angle of shearing resistance representing the maximum state of joint opening. Instability on slopes below this angle is associated with a regolith cover (grade V and VI) and not with material exhibiting rock structure. The concept of a residual angle in a completely jointed rock mass warrants further investigation.

Cut slopes in Wellington City are too young to indicate the sequence of development after the stable rock mass angle has been attained. However, natural slopes outside the city show that regolith of grade V and VI properties is present on slopes of 30° to 40°. In the urban area, regolith-covered slopes show widespread signs of instability on slopes in excess of about 36°. This condition may reflect degradation of the vegetation cover, slope modification or longer term climatic change. McConchie (1977) has demonstrated that in high magnitude rainfall events the greywacke-derived regolith is stable only at angles less than 19°. These two limiting angles (36° and 19°) correspond approximately to $\emptyset'$ and $0.5 \, \emptyset'$ for greywacke-derived regolith and suggest that the mechanisms observed by Skempton (1953) are controlling regolith stability (see Section 4.2 FLUVIAL DOWNCUTTING).

Skempton (1964), in discussing the significance of peak and residual strength in fissured clay, also noted the progressive loss in strength experienced by cut faces. By back analysis of landslides he noted that by 20 to 30 years after excavation, 60% of the strength difference between peak and residual values had been lost. After 50 years this had reduced to 80% and natural slopes recorded only residual strength.

4.3.2   REGOLITH SLOPE CHANGES

Carson (1969) presents a four stage model relating sequential instability to weathering-induced changes in rock-derived hillslope regolith (Table 4.6). The mechanisms of failure envisaged are similar to those proposed by Skempton (1953) (see Section 4.2 FLUVIAL DOWNCUTTING) but critical conditions are developed as a result of weathering-related property changes rather than landform development.

The first stage of instability occurs when joint development destroys rock cohesion sufficiently to reduce strength to the level of the prevailing shear stress. Talus derived from the ensuing failure of the rock face characterises stage 2 of the model and has a critical angle close to its angle of repose. In the absence of undercutting, this material will remain stable until its properties change. Because of the steep slope angles and high void ratios of jointed-rock and talus, these materials have efficient drainage. Positive porewater pressures are therefore unlikely to develop. For example, Pierson (1982) has measured hydraulic conductivities in excess of 10 000 cm/h in uncontaminated talus material. Consequently, in the absence of cohesion, the critical angles for jointed-rock and talus correspond to $\mathscr{G}'$ (compact) and $\mathscr{G}'$ (loose) respectively.

Table 4.6 **Four Stage Instability Model Based on Weathering-related Changes** (after Carson 1969)

Stages	1 Rock Face	2 Talus	3 Taluvium	4 Colluvium	
Threshold slope angle (β_c)	43–35°	33–38°	25–28°	sandy clay	19–20° 8–11°
Critical parameter	$\beta_c = \mathscr{G}'$ compact	$\beta_c = \mathscr{G}'$ loose	$\beta_c = 0.5\ \mathscr{G}'$	$\beta_c = 0.5\ \mathscr{G}'$	

Given time, talus will gradually change to taluvium – a mixture of rock debris and finer particles – represented in the model by stage 3. Carson (1969) attributes this change to rock breakdown. However, the development of taluvium may also be produced by the incorporation of airfall deposits such as loess (Pierson 1982) and volcanic tephra. The incorporation of wind-blown silt into a talus lattice is particularly common in mountainous areas during periglacial conditions. Braided river beds close to debris-clad valley-sides provide the optimum condition for the accumulation of fine particles within talus slopes.

As both the accumulation of fine particles within talus and talus formation may be episodic processes, many debris slopes consist of deposits of alternating texture rather than of uniform mixtures of taluvium. The hydrological properties of such slope deposits vary dramatically and thus the Carson model represents only the most simple situation. For example, Pierson (1982) describes three types of talus slope with hydraulic conductivities of 10 000 cm/h, 100 to 3000 cm/h and 10 to 100 cm/h, depending on the texture and stratigraphy of the deposits.

Augmentation of talus by fine particles increases total surface area of particle contact and consequently increases the angle of internal friction. The presence of fine material may also encourage aggregation of particles which can also lead to an increase in the angle of internal friction. Yee and Harr (1977) found that under a confining stress of 500 kPa, aggregation accounted for an increase in $\emptyset'$ of 9.5° to 11° over $\emptyset'$ of the soil in its disaggregated state. The additional strength was attributed to the greater angularity and rougher surface of the aggregates compared to the constituent primary particles.

With an increase in the proportion of fine particles, void ratio decreases, drainage potential drops and the likelihood of developing positive porewater pressure consequently increases. Despite the increase in the angle of internal friction of this material, the changes to hydrological properties have an overriding influence on potential stability. Carson argues that because these changes take place abruptly rather than continuously, a discrete state of instability may also be attained during the development of taluvium. The critical angle for one such state can be identified by both geomorphic evidence and stability analysis. Assuming the conditions outlined by Skempton and DeLory (1957), the critical angle for this state is approximately one-half the angle of internal friction.

Similarly, if critical conditions are not met before the taluvium weathers to colluvium, which characterises stage 4 of the model, then a new critical angle is achieved with respect to the properties of the colluvium.

The degree to which the sequence in the Carson model is achieved, the rate of change between instability stages and the value of critical angles (Table 4.6) will vary from place to place. The variation will be a function of lithology, rates of supply and removal and climatic conditions.

4.4   CHANGES OF REGOLITH DEPTH

The regolith is a thin, surficial cover of unconsolidated material commonly between 0.5 and 2 m in depth but in places attaining depths of up to 40 m. Compared to bedrock failures, regolith landslides are generally shallow movements with low depth/length ratios. Side and end stresses on regolith slides are therefore considered to be insignificant compared to those on the potential shear plane. This permits a form of infinite, limiting equilibrium analysis to be used in the representation of strength and shear stress conditions (Henkel and Skempton 1954). Because the rate of regolith formation is much less than the rate of change of climatic phenomena, it can also be assumed that slopes which are gentle enough to retain a regolith will at some time experience complete saturation. Thus in assessing long-term natural slope stability it is appropriate to invoke the most extreme transient conditions, represented by a water-table at the slope surface.

Employing the above assumptions and assuming also hydrostatic as opposed to artesian porewater pressure, the conditions of failure can be represented by:

$$\frac{s}{T} = \frac{c' + z.\cos^2 \quad (\gamma - \gamma_w)\tan \beta'}{\gamma.z.\sin\beta.\cos\beta} = 1.0 \qquad (4.2)$$

where s is shear strength
     T is shear stress
     c' is effective cohesion
     β' is angle of internal friction
     z is vertical depth of slide
     γ is total unit weight of material, and
     γ_w is unit weight of water

Regolith landslides are the most common form of slope movement and are nearly always associated with extremely wet weather conditions. Thus variation in the extent of saturation provides the best explanation for their day-to-day occurrence. However, other more passive factors are required to explain the long-term development of susceptibility to water table fluctuation.

One factor that changes as a result of both weathering and geomorphic processes is regolith depth. Critical depth (Z_c) at the time of failure is derived from equation 4.2 and is represented as:

$$Z_c = \frac{\dfrac{c' \sec^2 \beta}{\gamma}}{\tan \beta - \dfrac{(\gamma - \gamma_w)}{\gamma}\tan \phi'} \qquad (4.3)$$

In the following models, attainment of critical depth is considered the most appropriate explanation for both the onset of instability and the distribution of landslides throughout the terrain.

4.4.1 DEVELOPMENT OF CRITICAL DEPTH

Two concepts are important in this discussion. One is the role of regolith depth as a constraint to the development of instability on natural slopes. The other is the concept of a steady-state depth (Kirkby 1971) and how it relates to critical depth.

On natural slopes the very presence of regolith suggests that limiting conditions are not met over a considerable period of time. In other words, empirical boundary conditions exist for equation 4.2. Assuming through time a constant triggering regime and a constant slope angle, then boundary conditions must be represented by either the properties or the thickness of the regolith. However, some stable regolith consists of completely weathered material made up of secondary minerals in equilibrium with contemporary climatic conditions. The stability of this material thus points to the remaining factor, regolith thickness, as being the critical condition in the development of instability, at least in some circumstances.

Figure 4.7 shows the relationship between stable and unstable slopes with respect to slope angle and regolith thickness for granite-derived soils in Aichi Prefecture, Japan. The curved line represents the theoretical relationship between Z_c and slope angle when $c' = 7$ kPa and $\vartheta = 30°$. This diagram suggests that the regolith is stable until it reaches a depth close to that predicted by equation 4.3. As there is only one instance where the regolith is appreciably thicker than Z_c, it also appears that regolith thickness is determined by shallow sliding under saturated conditions.

Iida and Okunishi (1983) note that upper and lower limiting slope angles for sliding exist in a number of areas. They argue that at angles below the lower limit there is insufficient shear stress to promote sliding and that above the upper limit the regolith is too thin to allow this type of movement. The influence of slope angle on regolith thickness is through its effect on the rate of slow denudational processes such as slope wash and soil creep. Change in regolith thickness is also dependent on the rate at which the weathering front descends into the bedrock. The relationship of these factors is expressed as follows:

$$dz/dt = aVw - V'$$ (4.4)

where z is the thickness of the regolith,
Vw is the rate of descent of the weathering front,
a is the coefficient of expansion of material subject to weathering and,
V' is the rate of removal (positive) or accumulation (negative) by slow denudational processes

Theoretically, a steady-state thickness may be attained when aVw is balanced by removal (V'). This condition is represented in Figure 4.8 where Vw is shown as decreasing logarithmically with time (or depth as suggested by Ahnert 1966). There are a number of other expressions representing the relationships between Vw and depth and these can be found in geomorphology texts such as Carson and Kirkby (1972). In Figure 4.8, the rate of removal by slow denudational processes (V') is shown as being inversely related to Vw and increasing

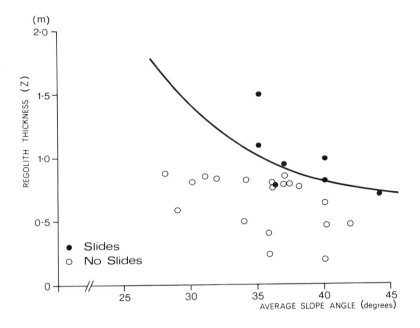

Figure 4.7   **The relationship between regolith thickness and slope angle for stable and unstable slopes.** The curve represents theoretically critical conditions (Iida and Okunishi 1983)

with time. This assumes that with time weathering will continue to reduce the strength of the regolith at a rate proportional to the production of regolith at the weathering front. Thus until time T_1 is reached, removal becomes easier with time. Whether actual removal follows this trend depends on whether applied stress or regolith strength is the principal constraint to denudation.

In Figure 4.8, T_1 is defined by the time taken to reach a balance between Vw and V' and represents the duration of the primary regolith building phase and indirectly the depth of the steady-state regolith. T_1 will vary with the form of the lines chosen to represent the rates of the two processes shown in Figure 4.8. However, after T_1 has been achieved, the rate of removal may temporarily exceed the descent of the weathering front. In this case, the regolith will be reduced, and consequently the rate of weathering will increase.

137

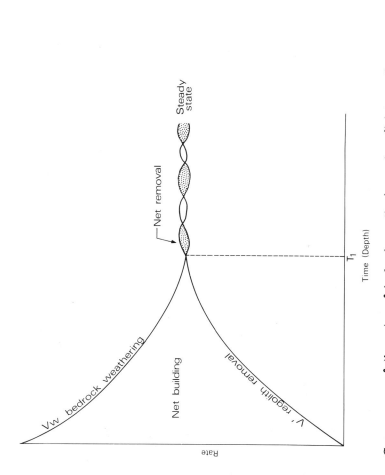

Figure 4.8 Convergence of the rates of bedrock weathering and surficial regolith removal by slow denudational processes to produce a steady-state regolith thickness

Correspondingly, should V' exceed Vw, surface lowering will begin to catch up with the descent of the weathering front, denudational agents will encounter progressively less weathered material and their rate of removal will consequently be reduced. Thus a steady-state or quasi-equilibrium thickness is theoretically possible as originally discussed by Ahnert 1967.

Many of the assumptions in the concept of a steady-state thickness are shown by Young (1972) to be more or less tenable depending on the form of the slope and other environmental conditions.

The theoretical development of regolith stability given above ignores the input of exogenous airfall material such as loess and volcanic tephra. In the case of volcanic tephra, thousands of square kilometres of the topography can be mantled in a period of a few days. In many cases the vegetation cover initially destroyed by the event can re-establish itself before much of the mantle has been eroded. The tephra deposits become secured close to their critical depth for the level of saturation experienced during the period of vegetation recolonisation. As time passes, the occurrence of extreme weather conditions will trigger sliding of the most unstable areas. Continued episodes of mass movement will require more extreme triggering events until the residual mantle is left only on gentle slopes where complete saturation has no effect. The remaining mantle can be considered to be in equilibrium with the prevailing exogenic regime.

In such circumstances periods of high erosion observed in the geological record may be simply a response to mantling by airfall deposits and not to a change in climate or other environmental conditions. The frequency of such eruptions may be high in terms of the geological time scale. For example, there have been 16 large eruptions of pumiceous tephra from the Taupo volcanic centre, New Zealand, within the last 40 000 years varying in volume from < 1 to 70 km^3. About 1800 years BP one such eruption mantled over 20 000 km^2 of the North Island of New Zealand to a depth of over 1 m (Dibble and Neall 1984).

Iida and Okunishi (1983) make use of the models outlined above to demonstrate the applicability of critical thickness and steady-state thickness in explaining the occurrence of landslides in Aichi Prefecture, Japan. In **Figure 4.9,** they indicate the integration of these models and the inclusion of landsliding itself as a denudational process reveals that only parts of the slope (between 20° and 50°) could possibly produce landslides. Outside this range the steady-state thickness cannot approach the depth required to produce failure, even under saturated conditions.

Because slope angle, catchment area, curvature, and microclimate affect both weathering rates and denudational processes, certain localised parts of a hillslope develop cri-

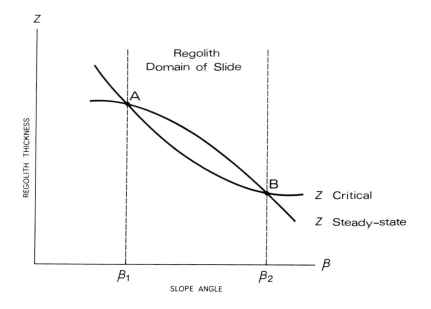

Figure 4.9   Hypothetical relationship between critical regolith thickness and steady-state thickness as functions of slope angle (Iida and Okunishi 1983)

tical regolith thickness more quickly and more frequently than do others.

4.4.2 LOCALISED INSTABILITY

Regolith landslides are commonly found to be preferentially located on certain parts of the terrain (Crozier et al. 1980). Two broad patterns of association occur (**Table 4.7 and Figures 4.10a and b**):

1. association with convergent, surface slope form (swales, dells, depressions, drainage lines) or with convergent subsurface slope form;
2. lack of association with convergent forms.

In the first case, the critical locational factor is often considered to be the concentration of water and the development of critical porewater pressure (Pierson 1977). This implies that the regolith on other slope forms would also

Table 4.7   Distribution of Regolith Landslides with respect to Slope Form (figures as percentage of all landslides)

Locality	Divergent (upper)	Planar Middle	Planar Lower	Convergent	Number of slides
1 Pakaraka	48	28	21	3	109
2 Notown	58	46	23	-	127
3 South Auckland	46	36	18		120
4 Omoto	8	11	23	58	277
5 Stokes Valley	-	-	-	60	78
6 Amakusa	8		7	85	273
7 Seirenji	3		2	95	292
8 Nishimikawa	9		6	85	455

Source: 1   Crozier et al. 1982
          2, 4   O'Loughlin and Gage 1975
          3   Selby 1976
          5   McConchie 1977
          6, 7, 8   Tsukamoto et al. 1982

fail if it could accumulate sufficient water. Thus during a climatic triggering event critical porewater pressures are achieved only within convergent forms. It also follows that in an extreme climatic triggering event, where saturation is achieved over most of the terrain, the preferential location with respect to convergent forms may be obscured as landsliding takes place on other saturated slope forms. However, the initial landslides occurring in such an event would still take place within convergent forms and only as the event progresses will the locus of activity shift. If there is no shift in the locus of activity under widespread conditions of saturation then the principal locational factor is probably depth of regolith rather than porewater pressure. This is because convergent forms often act as places of regolith accumulation compared with planar and divergent slope forms.

In the second case, where the pattern of occurrence is not clearly associated with convergent forms, some factors other than porewater pressure are operating as locational controls. These are commonly found to be slope angle, vegetation cover, regolith strength, slope aspect or condition of the regolith resulting from previous erosional activity.

The importance of convergent subsurface bedrock forms as the locus for debris slide activity has been recently highlighted by a number of geomorphologists working in the Pacific Northwest of the United States of America (Pierson 1977; Dietrich and Dunne 1978; Lehre 1981). All of these authors

0.5 km

Figure 4.10a   Debris slides and debris flows confined to zero-order basins from the December 1976 rainstorm of up to 350 mm in 24 h, western Hutt Hills, Wellington, New Zealand (Photo: Aerial Surveys Ltd, Nelson)

Figure 4.10b Upper slope initiation of debris slides and debris flows resulting from the 1977 rainstorm of 965 mm in 72 h, Kiwi Valley area, Wairoa, New Zealand (Photo: Hawke's Bay Catchment Board)

favour the following cyclic model for the occurrence of instability in U-shaped bedrock depressions containing soil (taluvial) wedges (Dietrich et al. 1982). Three principal stages are recognised in their 'Pacific Northwest Model' (**Figure 4.11**):

1. production of a U-shaped bedrock hollow by landsliding. This implies bedrock failure or scouring of bedrock by a debris slide;

2. infilling of the bedrock hollow. Material on the oversteepened perimeter of the bedrock hollow weathers quickly and is transported by slope-wash, splash or small-scale mass movement onto the floor of the hollow. Rates of sediment movement initiated by the hollow are thus higher than they are on undisturbed adjacent slopes. There are two phases of infilling:

   (a) as the bedrock-floored hollow concentrates overland flow, sediment initially arriving in the hollow will be sorted to produce basal layers of coarse residue. This phase is estimated to last about 100 years. As coarse layers accumulate, saturated overland flow becomes progressively less frequent in favour of throughflow;

   (b) in the second phase of infilling, throughflow has ascendancy over saturated overland flow and poorly sorted sediment, similar to that of the surrounding hillslope, begins to fill the hollow. Complete filling is estimated to take between 1000 and 10 000 years.

   The resultant soil wedge is commonly found within a topographic swale – referred to as a zero-order basin by Tsukamoto et al. (1982) (**Figure 4.12**). However, not all swales contain soil wedges and soil wedges may also be found on planar slopes (Marron in press) as well as on topographic rises (Eyles et al. 1978) (**Figures 4.13 and 4.14**). Marron (in press) notes that some bedrock hollows have concave upward longitudinal profiles ranging in gradient between 11° and 27°. **Figure 4.15** shows the depth/surface angle relationships for both regolith wedges and the surrounding hillslope regolith for two localities. A number of authors (Bryan 1940; Budel 1944; Cotton and Te Punga 1955; Beaty 1959; Bunting 1964; Dietrich et al. 1982; Tsukamoto et al. 1982) have discussed the origin and significance of regolith-filled bedrock hollows. Invariably these features are described as being contiguous with the drainage network and located headward of first-order channels or occupying positions similar to first-order channels;

3. the evacuation of the regolith wedge by debris

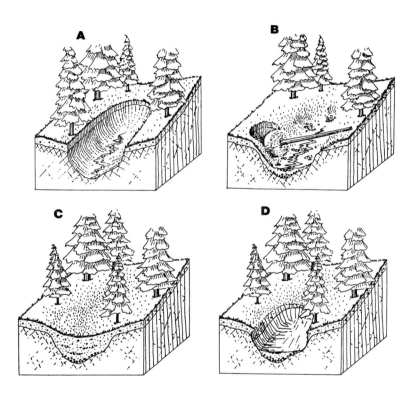

Figure 4.11   The Pacific Northwest model for the origin and
evolution of bedrock hollows (after Dietrich
et al. 1982):
A: bedrock landslide produces initial hollow
B: peripheral debris fills hollow and fluvial
sediment sorting takes place in hollow
C: filled hollow as a site for concentrated
subsurface flow and potential debris slide
D: evacuation by debris slide

Figure 4.12   Topographic swales ('zero-order' drainage basins)
extending to ridge tops on steeply dissected
greywacke terrain, Wellington, New Zealand
(Photo: Professor D W McKenzie)

slides.   Figure 4.15 shows that under saturated conditions,   regolith   wedges   are   closer   to   theoretical
critical   conditions   of   slope   angle   and   thickness   than
is   the   surrounding   hillslope   regolith.   The   critical
thickness   line   in   this   figure   is   based   on   surface
slope   and   may   overestimate   instability   because   the
bedrock   gradient   is   generally   lower   than   the   surface
slope.   The   susceptibility   to   failure   of   material   in
hillslope   depressions   has   been   demonstrated   by   Sidle

Figure 4.13 'Shaved' contact of a colluvially filled bedrock depression, Wellington, New Zealand
(Photo: Professor D W McKenzie)

Figure 4.14 A 'v'-shaped colluvially filled depression in faulted river gravels, Wellington, New Zealand (Photo: Professor D W McKenzie)

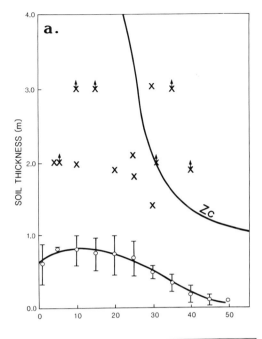

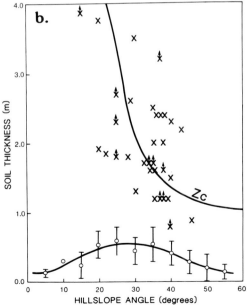

HILLSLOPE ANGLE (degrees)

Figure 4.15

Relationship between thickness of the regolith and slope angle for bedrock hollows and surrounding hillslopes (after Dietrich et al. 1982).

Data from Rock Creek, coastal Oregon (a) and Clearwater Basin, coastal Washington (b). The circles represent means and the bars represent standard deviations for the depth of hillslope regolith. The crosses represent regolith depth in the hollows and partial depth if attached to an arrow. The line Zc is the theoretical critical regolith depth under saturated conditions when $c = 8$ kPa; $\theta = 32°$ and $\gamma = 2$ g/cm^3

149

and Swanston (1982). They showed that an increase in porewater pressure from zero to 2.2 kPa was sufficient to reduce the factor of safety in one depression from 1.49 to 1.00 and hence to produce failure.

The most contentious aspect of the Pacific Northwest Model is the suggestion that the initial bedrock hollows are formed by landslides (Dietrich et al. 1982; Lehre 1982). The argument for this is based on contemporary environmental conditions but these appear unable to explain satisfactorily all the characteristics of these phenomena. For example, Dietrich et al. state: 'deep-seated landslides incorporating significant amounts of saprolite or bedrock are rare. Instead most landslides involve just the soil mantle and commonly emanate from wedges'. Marron (in press) reaffirms the suggestions that landslides produce the bedrock hollows by stating that: 'Small landslides are probably responsible for the creation of bedrock hollows and the partial emptying of colluvium-filled hollows.' Although she produces textural evidence for partial emptying, no evidence is given for complete emptying or, more significantly, bedrock sliding. She further observes that: 'Shallow landslides that partially empty colluvium-filled bedrock hollows are common in steep forested watersheds in the Pacific Northwest and are a likely mechanism for truncation of the observed buried soils.'

The evidence offered suggests that present-day landslides of a size and shape compatible with bedrock hollows are principally regolith phenomena in steep forested watersheds. Where then are the landslides which remove the appropriate volume and configuration of bedrock to form bedrock hollows? If they are not in evidence today, it is possible they may have occurred under a former morphogenic regime, one of periglacial characteristics. This is a reasonable assumption in view of the latitude of the Pacific Northwest and northern California mountains and the age of some soil wedges. Marron (in press) provides a date of 7000 years BP for one of the least weathered wedges in her study area in northern California, suggesting that many wedges will be older than this. The hollow itself will, of course, pre-date the infilling of the wedge. There is also evidence of limited glacial and periglacial activity during the late Pleistocene in areas of the Pacific Northwest where bedrock hollows have been described (Swanson and James 1975). Argument for a landslide origin for hollows, based on shape and location of present-day landslides, is somewhat circular and clearly inappropriate in view of the age of the features.

To discount a fluvial origin for bedrock hollows on the grounds that no alluvial sediments occur in regolith wedges is also not convincing because alluvial sediments, in any case, are absent or scarce in most first-order steep mountain tributaries. Contiguity of bedrock hollows with the drainage network and the 'V' rather than U-shaped cross-sections of many hollows suggests that water pathways rather than surfaces of rupture may be influential in the formation of many hollows. Bunting (1964) has studied soil-filled, bedrock hollows occupying topographic depressions of a similar configuration (but of a gentler slope) to those described in the Pacific Northwest. These features, referred to as 'percolines', are not associated with landslides and are ascribed to the weathering and eluviation effect of concentrated subsoil water flow. They are considered to be the precursors of first-order stream development. Channel development by gullying has been observed in these features in unforested slopes in the western Pennines.

4.5   ALTERNATING EXOGENIC REGIMES

In the sense used by Tricart and Cailleux (1972), a morphogenic regime is the system within which exogenic and endogenic processes interrelate to produce a characteristic landform response. They define climatic geomorphology, which treats climate as the ultimate control of exogenic processes, as: 'the study of relief forms as determined by climate'. Because of the large amounts of time required in landform evolution and hence the opportunity for morphogenic processes to change within that period, climatic geomorphology is a complex undertaking. It depends for its success on the correlation of relief forms with particular morphogenic systems. Unfortunately an existing landform may have experienced several different climatic and process regimes during its development. Isolating the dominant regime or the integrated influence of a number of regimes is extremely difficult if not impossible in many cases.

In recognition of these problems, the term 'exogenic regime' is used here to refer to the system of geomorphic processes, their type and intensity, which come about as a result of the reaction of climate with vegetation, soils, and topography. The regime may produce or tend to produce a characteristic landform response. The important difference between this concept and Tricart and Cailleux's definition of climatic geomorphology is that this approach is not dependent on the existence of particular relief forms nor does it focus on those forms. The concept of an exogenic regime is useful in slope stability studies because a shift to a new regime may unfavourably affect the passive stability factors without producing any discernible relief response. Change of exogenic regimes may be brought about by climatic changes: mantling of the topography by airfall deposits, and by deforestation or other disturbance to the vegetation cover.

4.5.1   PLEISTOCENE/POST PLEISTOCENE

To return to the question of bedrock hollows, Cotton and Te Punga (1955) have reviewed a number of models for their formation, many of which depend on the alternation of exogenic regimes. One model involves modification of first-order, headwater channels by mass movement processes. Gully filling takes place when vegetation is suppressed and regolith activity is promoted by a new exogenic regime. For example, the fine-textured (feral) relief of Wellington, New Zealand, is thought to have been developed by fluvial activity during an interstadial or succession of interstadial periods, such activity being subsequently subdued, initially in the upper catchment, with the onset of a periglacial regime (Cotton 1958). Palaeobotanical evidence (Stevens 1974) indicates that the upper tree limit in the region during the peak of the last glacial (20 000 years BP) was somewhere close to the present-day sea-level; that is, about 1070 m lower than it is at present. Under present-day forested conditions, however, regolith movement is much less extensive than it was in the past and takes place only episodically, principally in bedrock hollows during extreme events.

Another variant on the alternating exogenic regime origin for regolith-filled bedrock hollows (put forward by Cotton and Te Punga (1955)) involves the extension rather than modification of first-order channels by the corrosive action of periglacial solifluxion. In the post Pleistocene period, re-establishment of the forest cover has stabilised the periglacially produced debris within the hollows. Much of the terrain studied by Cotton and Te Punga has been deforested in the last 100 years and recent landslide episodes in this area have evacuated many regolith wedges (Eyles et al. 1978). The result of these landslides has been an extension of total first-order channel length and a corresponding increase in drainage density. At least in the last 10 years there has been little evidence that infilling will re-assert itself over channel processes that are now operating in former bedrock hollows.

There is good evidence in the steep, hard rock terrain of the Wellington region to indicate that the infilling of bedrock hollows (and possibly their formation) is a product of a periglacial exogenic regime. It also appears that evacuation by debris slides and development of channelisation characterises the present temperate exogenic regime. Deforestation has served to accelerate this process.

4.5.2   VARIATION IN CLIMATIC PATTERNS

Much of the currently severe erosion occurring in the New Zealand hill country and mountain land has been traditionally attributed to the destruction of the forest and the attendant land use activities. Cumberland (1947) sums up this view of New Zealand by stating:

Its cultural youth has been characterised to a large extent by the pioneer destruction of resources of a little known environment. Soil erosion is one of the more disastrous outcomes of this historical fact and it is a problem that has grown with the growth of the nation.

However, recent work in New Zealand (Crozier 1983; Grant 1983) has identified phases of high erosional activity which either pre-date human occupation of the country or which cannot be reasonably attributed to cultural activity.

In particular, Grant (1983), working in the Ruahine Range, identifies five erosional phases (**Figure 4.16**), four of which occurred in the absence of deer, opossums and other introduced wild animals. Furthermore, the first three phases took place before the major cultural impact of European settlement but within the period of Polynesian occupation. However, each phase of erosion, apart from the Waipawa phase, has been succeeded by a relatively long, soil-forming, tranquil phase despite human presence. Of greater significance is the measured increase in erosional activity since 1950 AD which has taken place during a period when there has been reduced cultural interference with the vegetation cover. **Figure 4.17** shows the rapid increase in erosion since this time.

Because of the lack of correlation between erosion rates and demonstrable, culturally-related disturbance and the fact that the most important sediment sources occur in already poorly vegetated areas, Grant (1983) has searched the climatic record for a possible cause of the enhanced erosional activity. He concludes that since the 1950's, wide-scale changes of atmospheric pressure and circulation have occurred in the south-west Pacific region. This has increased the incidence of warm, moist, northerly air over the country and has led to an increase in storm rainfall and heightened flood activity. These changes appear to have increased the incidence of landslides, enhanced sediment transport from steep, upper catchment slopes and promoted deposition in the middle catchment.

By analogy, Grant (1983) has attributed the earlier phases of erosion to similar, climatically-induced shifts in the exogenic regime. However, the upper catchment activity identified by Grant has been interpreted on the basis of variations in sediment storage within the middle catchment. In such cases, the time lag between upper-catchment slope erosion and sediment transport within the main channel needs to be taken into account. Madej (1982), for example, has shown that in a 23 km^2 catchment in western Washington, sediment

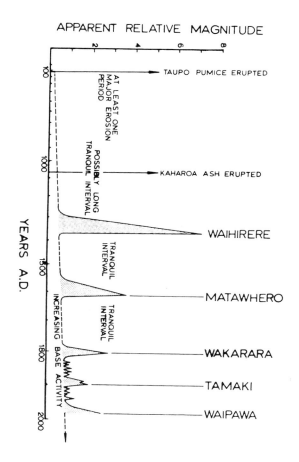

Figure 4.16   The chronology and relative magnitudes of erosion phases within the Waipawa Valley, Ruahine Range, New Zealand (Grant 1983)

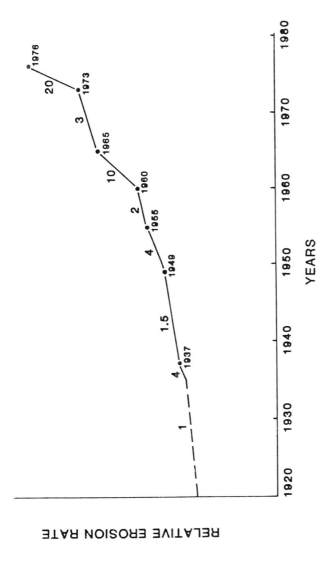

Figure 4.17 Relative rates of erosion, sediment transport and deposition in the channel of the upper Waipawa River since 1920. The trend curve was produced by plotting each periodic gradient using its rate index as a ratio to 10, for example, the 1973–1976 rate was five times that of 1949–1955 (Grant 1983)

placed in the channel by present disturbances will take 20 to 40 years to be removed.

The work of Bergstrom and Schumm (1981) indicates that inferring hydrological events from episodic deposition and erosion within stream systems can be highly misleading. They state that:

> Episodic behaviour is not controlled by hydrological events but by geomorphic threshold conditions. Depositional and erosional processes tend to build landforms towards critically unstable forms (thresholds) which will eventually fail causing a series of erosional and depositional events...Such behaviour may be intrinsic to the fluvial system...

4.5.3 VEGETATION CHANGES

The vegetation cover is one of the most influential and sensitive elements of the exogenic system. Even minor environmental pressures may result in change to its structure and composition, particularly if it is already under stress. Communities located close to their altitudinal or latitudinal growing limits are thus often the first to respond to pressure. These pressures can be brought about by climatic change, disease, fire, synecological competition or browsing activity (Dansereau 1957) but the most radical changes result from accidental or intentional destruction by man. Forest covers have been harvested, destroyed, or converted to pasture in relatively short periods of time, over vast areas of the earth.

The stability of hill country and mountain land soils has proved very sensitive to changes in the vegetation cover. In the western United States, for example, forest clearance has accelerated debris avalanche erosion by a factor of two to four, and debris avalanche erosion related to the development of forestry roads has increased 25 to 340 times, compared with its level in forested areas (Swanston 1976). Debris torrents, often triggered by debris avalanches, also increase in response to timber harvesting activity. Swanston (1977) found that the density of debris torrents (events/km^2/y) on deforested areas in Oregon increased 4.5 to 8.8 times, compared to forested areas. In the same region, Swanson and Dyrness (1975) recorded a 2.8 times increase in the volume of material moved by debris slides on clearcut compared to forested slopes, over a 20 year period. In the Biratori and Urahoro areas of Japan, Fujihara (1978) has reported that clearcut areas have eight to nine times more landslides than do forest areas on similar terrain and that the eroded area is seven to eleven times greater. On greywacke hill country in the North Island of New Zealand, Selby (1976) estimates that conversion from forest to pasture increases the probability of occurrence of a landslide event of a given magnitude by about three times. O'Loughlin and Pearce (1976) have studied the effect of clear-felling and slash-burning podocarp-hardwood forest on steep, sandstone hill country in Westland, New Zealand. Forest clearance increased debris slide/debris avalanche erosion from an estimated long-term 'natural' rate of 100 $m^3/km^2/y$ to an average of 4000 $m^3/km^2/y$, over a three year period.

Together with similar results from other parts of the world these data indicate that, in the short term, deforestation greatly accelerates mass movement erosion. However, four points should be kept in mind in interpreting and applying such results. First, large landslides can occur under healthy, mature forest, albeit at a low frequency. Thus re-afforestation programmes are not likely to prevent landslide activity completely. Second, as indicated by Grant (1983), although an area may have suffered considerable degradation of its vegetation cover, other factors may now be controlling trends in erosion rates. Third, the vegetation cover may actually become a destabilising factor under certain conditions, particularly on steep jointed-rock, in areas exposed to high winds and where slope movements are deep and fluid (see Sections 3.3.3 VARIATION IN WEIGHT and 3.3.7 JOINT PROCESSES - Root wedging). Fourth, the fact that landslides in forested areas are usually larger (although less frequent) than those in deforested areas suggests that, over a very long term, there may not be a great difference in the volume of material lost under different vegetation covers.

The stabilising influences of forest cover that may be lost or reduced as a consequence of deforestation are:

1. surcharge: on most forest soils (often with little or no cohesion and only residual internal friction) surcharge brought about by the presence of forest cover tends to increase slope resistance. However, O'Loughlin (1974) and Wu et al. (1979) estimate that surcharge provides a normal stress of only 1 to 5 kPa which is small compared to the stresses involved in the soil regolith above a potential shear plane;

2. hydrological effect: vegetation can remove large amounts of soil water by evapotranspiration and to a lesser extent can reduce the precipitation supply by interception (Hallin 1967). In some cases, these influences are sufficient to control the level of perched water-tables, as is evident by their appreciable rise in some areas after deforestation (Conacher 1974; Stone et al. 1978). Compared to deforested areas, regolith under forests maintains relatively low antecedent moisture conditions and has a shorter period of seasonal saturation. This may lower the frequency with which rainstorms trigger

mass movement. However, this influence may be obscured in areas subject only to high magnitude rainstorms;

3. <u>organic influence on soils</u>: organic debris and root pathways tend to promote or maintain freely draining soils. Litter layers prevent the sealing of the surface by raindrop impact and clay/organic molecules maintain soil structure and encourage biological activity. Under most rainfall conditions, the hydraulic conductivity of forest soils prevents critical accumulation of groundwater. However, under very high rainfall intensities, the relatively high infiltration rates of forest soils may increase their susceptibility to mass movement, compared to soils with lower infiltration rates;

4. <u>microclimatic effect</u>: the forest microclimate prevents extreme variation of temperature, humidity, windspeed and soil moisture within the forest. Consequently, high rates of soil expansion and contraction and the development of desiccation cracking are not common. Deforestation and consequent exposure of certain fine-textured soils can lead to desiccation cracking, loss of cohesion, and the development of localised subsurface erosion;

5. <u>mechanical reinforcement by tree roots</u>: tree roots provide the most important stabilising influence on forest soils and operate to reinforce the ground in three ways:

   (a) by attaching potentially unstable regolith to stable substrates;

   (b) by providing a matted network which offers lateral attachment near the surface; and

   (c) by providing a localised zone of reinforcement associated with individual trees, similar to the arching restraint provided by piles.

O'Loughlin and Ziemer's (1982) synthesis of a number of studies concludes that the additional strength provided by tree roots (artificial cohesion Δc) can be treated as an increment to effective cohesion (c') in equation 3.1, by an amount varying between 1 and 20 kPa. Both the size and number of roots are important factors in determining the magnitude of added strength. For example, from <u>in situ</u> shear tests on <u>Pinus contorta</u> roots in sandy soil, Ziemer (1981) calculated that increasing amounts of dry root biomass (kg/m^3) may increase soil strength as follows:

$$\text{soil strength} = 3.13 + 3.31 \text{ biomass} \qquad \textbf{(4.5)}$$

However, not all roots have the same influence on soil strength and a number of studies indicate that small roots

(< 25 mm diameter) are the most important (Burroughs and Thomas 1977; Ziemer and Swanston 1977; Ziemer 1981). The tensile strength of individual roots appears to be more important in slope resistance than is resistance generated at the root/soil interface. This is because, under applied stress, roots tend to break rather than pull out of the soil, although root slippage has been observed in some tests on saturated soils (Waldron and Dakessian 1981).

After deforestation, small roots lose their tensile strength by decay, at rates averaging between 300 and 500 kPa per month (O'Loughlin and Ziemer 1982). This loss of strength is often offset to some extent by recolonisation of the site. Taking into account both root decay and recolonisation rates, Ziemer (1981) showed that total root strength influence was at its lowest level seven years after logging.

The manner of deforestation and subsequent ground treatment will also affect the degree and rate of strength reduction that occurs on forest clearance. For example, forest fires of certain intensities appear to induce or enhance a hydrophobic condition in the soil which may promote surface runoff and thus prevent critical groundwater accumulation on affected sites (Wells 1981).

4.6   GEOMORPHIC HISTORY

Most of the instability models presented so far have involved some aspects of general geomorphic history. In the following two models, however, geomorphic history is used specifically to explain the variation in stability of different parts of the contemporary landscape. In the first model, rejuvenation of fluvially-dissected, deeply-weathered terrain leads to the unusual result of upper slope instability. In the second model, the nature and location of previous landslide activity dictates a shift in the locus of active instability.

4.6.1   DEEPLY WEATHERED TERRAIN

Crozier et al. (1981) have related the development of instability on deeply weathered volcanic rocks in Fiji to the geomorphic history of the area. During extremely intense cyclonic rainfall, debris avalanches and debris slides occur predominantly within deep, red, clay-rich regolith on the shoulders of slopes within lower catchment areas. The other type of regolith within these catchments is a shallow, brown, stony taluvium occupying slopes subject to fluvial under-cutting. Despite the steeper angle of this material it is relatively stable during intense rainstorms. Brown regolith slopes derive their stability largely from the gravel fraction which both enhances internal friction and increases permeability. Loss of strength which may result from the development of positive porewater pressures or from a reduction of apparent cohesion with high water content within the red clay-rich regolith, is therefore not experienced to the same degree in brown regolith. The shallowness of brown regolith also allows tree roots to penetrate the bedrock surface.

A model relating the distribution of regolith to geomorphic development and landform position within a drainage basin is presented in **Figure 4.18.** The sequence (a) through (d) illustrates idealised cross-sections of the main valley, from the headwaters to the lower catchment. Diagrams (a), (b) and (c) represent the sequence expected during one 'cycle' of hill country erosion while (d) may be interpreted as indicating a period of renewed stream incision and 'rejuvenation' of relief. Lithological differences could provide an alternative explanation for the two phase nature of the slopes in the lower catchment (d) but no evidence was found to support this possibility.

The differences in form and degree of weathering among (a), (b) and (c) are ultimately attributable to the increase in valley size and floodplain development which result from greater stream discharge and a longer period of fluvial activity being experienced in the lower parts of the catchment compared to the upper parts. This interpretation presupposes a headward extension of the drainage network and a decrease in stream gradient with time. A corollary of the model is that the landmass presently located at (c) would have experienced a sequence of development through (a) and (b) stages and, if (d) is interpreted as indicating an interruption to the erosion sequence, it would have been preceded by (a), (b) and (c) stages before 'rejuvenation' took place.

'Rejuvenation' appears to have been either too recent or too insubstantial to have been reflected in slope conditions further up the catchment. There is, however, a marked water-fall on many main channels immediately upstream of the floodplain which may represent some form of knickpoint associated with 'rejuvenation'.

The significance of valley widening and floodplain development to regolith formation lies in its influence on the frequency with which stream action is experienced directly at the toe of the slope. Diagrams (a) and (d) represent situations where stream contact with the slope prevents accumulation of slope deposits and maintains active colluvial transport, steep slopes and 'young' soils, characteristic of the strong brown regolith. Unlike the slopes represented in the early stages of stream dissection in Skempton's (1953) and Carson's (1969) models, the brown regolith slopes in (a) and (d) show no evidence of previous landsliding. Even during a cyclone event where over 900 mm of rain fell in 48 hours, these slopes produced relatively few landslides.

In contrast, slopes in diagrams (b) and (c) represent 'older' land surfaces where weathering has been allowed to

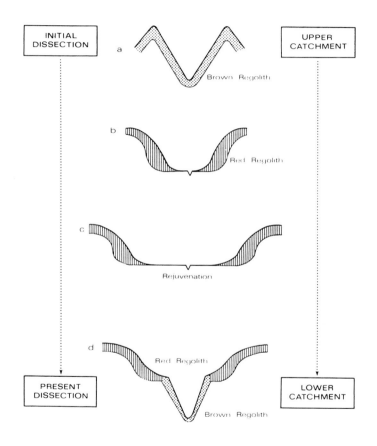

INITIAL
DISSECTION

UPPER
CATCHMENT

a

Brown Regolith

b

Red Regolith

c

Rejuvenation

d

Red Regolith

PRESENT
DISSECTION

LOWER
CATCHMENT

Brown Regolith

Figure 4.18   Model for the development and 'perching' of
unstable red regolith by weathering and rejuvenation of fluvial erosion, south coast Viti
Levu, Fiji (Crozier et al. 1981)

proceed without interference from stream-induced slope activity. This situation permits the formation of the deeply
weathered, red regolith which exhibited a high degree of
instability during the rainstorm. The increase in depth of
weathering depicted at the toe of the red regolith is partly
speculative in the case of (b) and (c) but was observed in a
number of cases in the situation depicted in (d). After
removal of the red regolith by landslide, a well defined
amphitheatre-like depression was revealed in the bedrock surface. This shape, in itself, would aid in the accumulation of
slope water and enhance the instability of the regolith during
a period of high rainfall. When these deeply weathered slopes
become perched as a result of stream incision (d), they are
left without lateral support and as a result are particularly
susceptible to failure.

4.6.2 PROGRESSIVE REGOLITH STRIPPING

The unusually wet winter of 1977 in the Wairarapa hill country
of New Zealand produced one of the most extensive regolith
landslide episodes ever recorded in the country. The landslides occurred in a distinctive distributional pattern,
strongly favouring the normally dry and 'sunny' upper parts of
northerly-facing slopes (**Figures 4.19; 4.20** and **Tables 4.8;
4.9**).
          Crozier et al. (1980) initially correlated the distribu-
tion with slope angle, rock type, vegetation cover, slope
drainage and rainfall distribution during the episode, but
none of these factors could explain the peculiar distribution
of unstable slopes. They concluded that aspect-related differences in regolith conditions were probably influencing the
distributional pattern.
          The region had experienced a large number of mass move-
ment episodes in the past (**Figure 4.21**), each one stripping
large areas of regolith from the underlying sedimentary rock.
Trustrum (1981) has determined that the cumulative eroded area
on average represents 39% of the land surface. In virtually
all cases, the landslides that occurred in 1977 were found to
originate within the remaining undisturbed regolith. Mapping
the distribution of undisturbed regolith showed that it is
located on the upper slopes, summits, and ridge crests and
that it is more extensive on northerly-facing slopes than it
is on other slope aspects.
          To some extent the aspect distribution of undisturbed
regolith is able to account for the distributional pattern of
landslides (**Figure 4.22**). Thus availability of susceptible
material, resulting from the history of regolith stripping,
broadly determined the location of landslides. In 1977,
undisturbed regolith on all aspects was susceptible to failure
but greater opportunity for failure existed on northerly-
facing slopes. Material formerly present on the steepest slo-

Figure 4.19   The preferred locations of landslides in
northerly-facing catchments and on northerly
slope aspects from winter of 1977, eastern
Wairarapa hill country, New Zealand (Photo:
Department of Lands and Survey)

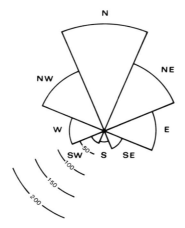

Figure 4.20   Landslide density (number/ha) with slope aspect for the 1977 episode, eastern Wairarapa hill country, New Zealand (Crozier et al. 1982)

SLIPPING EPISODES

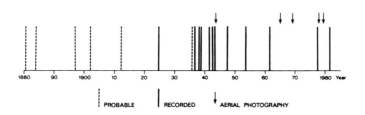

Figure 4.21   The chronology of landslide episodes since 1880, eastern Wairarapa hill country, New Zealand (Crozier 1983)

Table 4.8 Aspect Preference for Landslide Occurrence, During the 1977 Episode, Wairarapa, New Zealand (Crozier et al. 1980)

				Compass Octant					
	N	NE	NW	E	W	SE	SW	S	Total
Land area (ha)	242	259	301	395	305	253	285	220	2260
Area percent	10.7	11.5	13.3	17.5	13.5	11.2	12.6	9.7	100
Number of slips	547	404	407	426	206	97	75	44	2206
Slips percent	24.8	18.3	18.5	19.3	9.3	4.4	3.4	2.0	100
Slips per hectare	2.26	1.56	1.35	1.08	0.68	0.38	0.26	0.20	0.98

Table 4.9 Hillslope Position of Landslide Crowns; 1977 Landslide Episode, Wairarapa, New Zealand (Crozier et al. 1982)

	Lower third	Middle third	Upper third	Total
Number	336	681	1189	2206
Percent	15	31	54	100

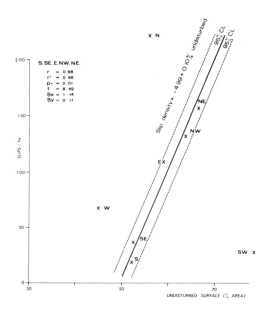

Figure 4.22   The relationship between the density of landslides in 1977 and the amount of undisturbed
regolith existing prior to 1977, within slope
aspect groups, eastern Wairarapa hill country,
New Zealand (Crozier et al. 1980)

pes and in hydrologically sensitive depressions and lower
slopes had been removed by earlier episodes. The remaining
regolith was thus left in the more resistant sites for
possible removal in subsequent but presumably more extreme
events. Thus a type of 'event-immunity' or 'event-resistance'
was acquired from successive exposure to periodic mass movement in an area of rapid regolith stripping.

Figure 4.22, however, indicates that in the Wairarapa
case, 'sunny' slopes experienced more landslides than could be
explained solely by the distribution of the deep, undisturbed
regolith. Variation in the shear strength of regolith on different slope aspects may be responsible for some of this
unexplained susceptibility to failure. Direct shear strength
tests at different water-contents (Owen 1981), as well as
other evidence, show that, at a given water content, regolith
on 'sunny' slopes is weaker than that on 'shady' slopes
(Figure 4.23). However, when the 'normal' aspect differences

in water content experienced in the field are superimposed on this diagram, 'shady' slopes are seen to be less resistant than are 'sunny' slopes **(Figure 4.24)**. The higher and more prolonged antecedent moisture conditions experienced by 'shady' as compared to 'sunny' slopes makes the former normally much more susceptible to triggering events. It is, therefore, not surprising that more extensive regolith stripping has taken place in the past on 'shady' slopes, thus leaving a greater extent of undisturbed regolith on 'sunny' slopes prior to the 1977 episode.

Analysis of the climatic conditions prevailing in the region in 1977 (Crozier et al. 1982) indicated that they were extremely unusual. In contrast to normal winter conditions, high rainfall totals, a large number of rain days, low incidence of northerly winds and low evapotranspiration rates, allowed antecedent moisture conditions on northerly-facing slopes to attain levels equivalent to those on 'shady' slopes thereby offering less resistance than shady slopes (Figure 4.24). Together with the greater availability of deep, undisturbed regolith, the upper, northerly-facing slopes were thus primed for failure during the rainfall events of the 1977 winter.

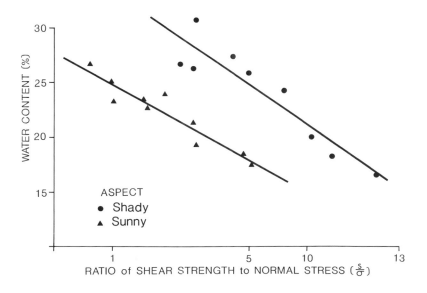

Figure 4.23   Variation of regolith strength with water content for 'sunny' and 'shady' aspect slopes (Owen 1981)

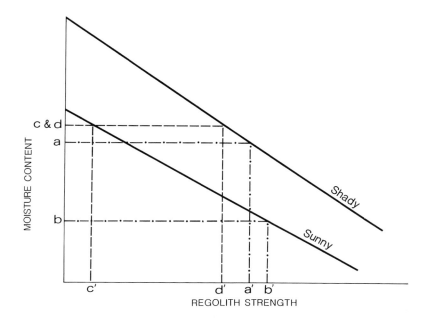

Figure 4.24   Strength of the regolith on 'sunny' and 'shady'
slopes during normal conditions (a = shady; b =
sunny) and the unusual conditions experienced
during the winter of 1977 (c = sunny; d =
shady), eastern Wairarapa hill country, New
Zealand (after Owen 1981)

CLIMATIC TRIGGERING $\mathbf{5}$
OF LANDSLIDE EPISODES

Much mass movement occurs and most regolith landslides are initiated during intense precipitation events or during lesser precipitation events associated with prolonged wet periods. The wet periods generally result from persistent antecedent rainfall, rapid snow and ice melt, or from the conservation of soil water by low evapotranspiration conditions.

When critical weather conditions are attained, landslides commonly occur in regional clusters including, in some cases, thousands of individual landslides distributed over hundreds of square kilometres (**Figure 5.1**) with varying densities (**Table 5.1**). The well-defined cellular pattern of many climatically-triggered landslide clusters is related to the isohyetal pattern of the associated weather systems (**Figure 5.2**). In many parts of the world low pressure cells, particularly tropical cyclones, are the major source of landslide-triggering rainstorms. Frontal convergence, deep troughs of low pressure and the orographic enhancement of moist, ascending air have also been responsible for triggering rainstorms.

In inherently unstable regions, major landslide episodes may occur on average once every five to six years (Crozier et al. 1982). On a broader scale, New Zealand, for example, may experience approximately one economically-significant landslide episode somewhere in the country almost every year (**Figure 5.3**). However, many landslide episodes in remote areas are probably never recorded whereas those occurring in populated areas may receive considerable attention. In Serra das Araras, Brazil, for example, a number of studies have been made of the devastating episode of 22-23 January 1967 which has been described as: 'laying waste to a greater land mass than ever recorded in the geological literature' (Jones 1973). In this event a total of 586 mm of rain fell in 48 hours over an area of 180 km^2 (Da Costa Nunes 1969). Another well-reported, rainfall-induced episode, occurred in Hong Kong in 1966 and is described as having produced the most disastrous mass movement ever recorded in the region (So 1971).

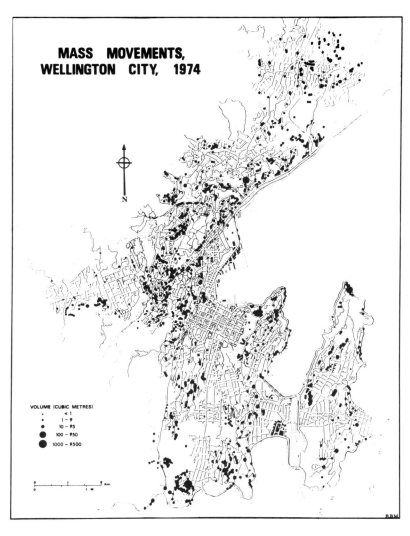

MASS MOVEMENTS, WELLINGTON CITY, 1974

VOLUME (CUBIC METRES)
. < 1
. 1 – 9
• 10 – 95
● 100 – 950
● 1000 – 9500

Figure 5.1   Shallow rock and debris slides occurring in the winter of 1974, Wellington City, New Zealand (Eyles et al. 1978)

Table 5.1   Comparison of Landslide Densities (number/km^2)
for Rainfall-triggered Episodes

(1)	Pakaraka Wairarapa, 1977	478	(7)	Wellington City Winter 1974	19
(2)	Wairarapa, 1977 (Regional mean)	98	(8)	Wellington City December 1976	16
(3)	Wainitubatolu April 1980	34	(9)	West Coast, SI Waimaungan gravels 1973-75	11
(4)	Tangoio, Hawkes Bay May 1971	31	(10)	West Coast, SI Old Man gravels 1973-75	10
(5)	Puriri, Hauraki 1981	31			
(6)	West Coast, SI Sandstone, 1973-75	19	(11)	Stokes Valley December 1976	6

Note: All data are from New Zealand localities except for (3) which is from Viti Levu, Fiji.

Source: 1 and 2, Crozier et al. (1982)
3, Crozier et al. (1981)
4, Eyles (1971)
5, Eyles and Eyles (1982)
6, 9 and 10, O'Loughlin and Pearce (1976)
7 and 8, Eyles et al. (1978)
11, McConchie (1977)

5.1   THE NATURE OF THRESHOLDS

The fact that some rainstorms produce landslides and others do not has prompted a search for the climatic parameter which will provide the most accurate indication of the triggering threshold. Ideally this parameter should be specific to the hydrometeorological event itself but because such measurements are not always available, standard, calendar-defined meteorological parameters are most frequently used.

Four different methods can be used to determine triggering thresholds:

1. theoretical models: where inputs of precipitation are considered to be distributed through the soil mass at previously determined rates and where derived quantities of water are presumed to alter the factor of safety;

2. empirical tests: where field experimentation under controlled conditions is used to generate failure;

3. spatial correlation of landslide occurrence and climatic events;

4. temporal correlation of landslide and climatic events.

171

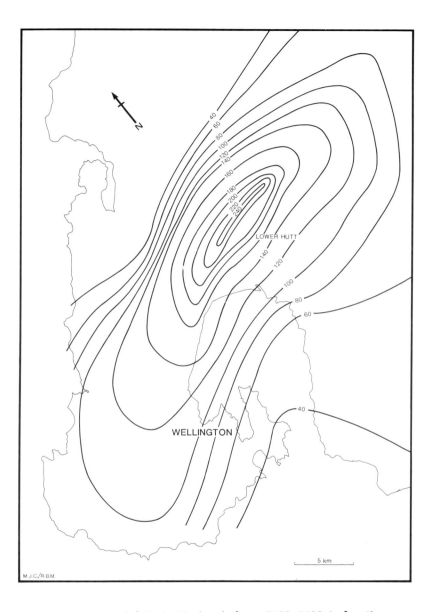

Figure 5.2   Rainfall depth (mm) from 0400-1400 h for the
              Hutt Valley/Wellington rainstorm, 20 December
              1976 (Bishop 1977)

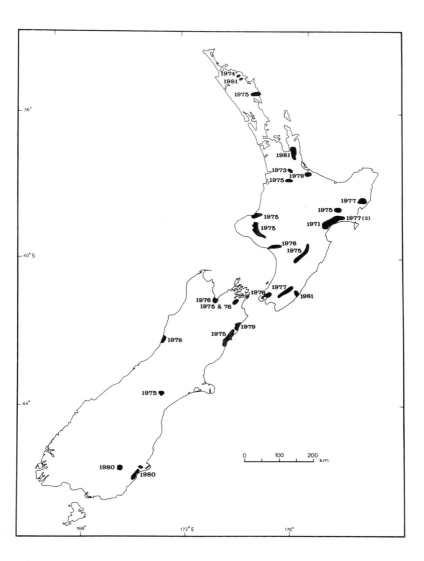

Figure 5.3   Landslide episodes producing damage, as recorded
                in head office files, Water and Soil Division,
                Ministry of Works and Development, New Zealand
                1970-1981 (Eyles and Eyles 1982)

173

The reliability of the threshold established by any of these methods depends on the type of climatic parameter chosen, size (and therefore homogeneity) of the area under consideration, completeness of the data base, and the way in which mass movement is defined. **Figure 5.4a** provides one example of spatial correlation as a means of establishing triggering thresholds. It also illustrates the increase in severity of landslide activity with increasing rainfall intensity (**Figure 5.4b**) and hence demonstrates the need to define the degree of mass movement activity to which any given threshold refers. The landslide episodes recorded in the San Dimas Experimental Forest illustrate a similar effect between different-sized rainstorms (**Table 5.2**). In the Coast Mountains of California, Lehre (1982) identified virtually the lower triggering limits when a 24-hour rainfall of 115 mm produced only two landslides; three months earlier, in the same locality, a 24-hour rainfall of 165 mm produced 29 landslides.

Table 5.2   **Severity of Landslide Erosion in Two Rainstorm Periods, San Dimas Experimental Forest, California** (Rice and Foggin 1971; Rice 1982)

Date	Amount (mm) (duration)	Return period (y)	Area eroded (%)	Erosion rate (m^3/ha)
November/ December 1966	282 (4 days) 157 (2 days)	9	1.7	21
January/ February 1969	802 (8 days)	32	5.5	298

Two main types of triggering threshold may be identified for any given level of mass movement:
1. the <u>minimum probability threshold</u> (PTn) below which the defined level of mass movement does not occur and above which it may occur under certain conditions;
2. the <u>maximum probability threshold</u> (PTx) above which the defined level of mass movement always occurs; that is, there is a 100% probability of occurrence whenever PTx is equalled or exceeded.

The range between PTn and PTx is referred to as the <u>probability margin</u>.

Because of the destructive nature of mass movement, triggering thresholds must be considered unique to the prefailure condition of the slope (Crozier 1972). As the prefailure condition may never be replicated, established thresholds have only limited predictive capability. For

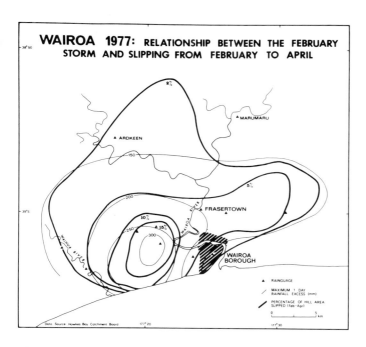

Figure 5.4a   Area eroded and rainfall depth during the 1977 Wairoa landslide episode (Eyles and Eyles 1982)

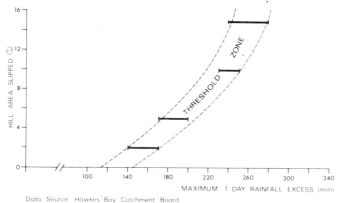

Figure 5.4b   Relationship between area eroded and rainfall depth for the episode in Figure 5.4a showing the probability margin (threshold zone) for defined levels of erosion (reproduced with permission of R J Eyles)

example, some slopes may acquire event-resistance; that is, the triggering threshold may increase with each successive event. This is generally associated with rapid stripping of regolith when the most susceptible sites fail in early events, leaving the more resistant sites to fail in subsequent events. Table 5.3 provides an example of event resistance. Prior to the 1974 event shown in this table, a period of about 10 years had elapsed since the area had experienced any major landslide activity. Although the relatively low landslide activity during the 1976 event is most probably the result of event-resistance, the lower antecedent rainfall of 1976 may have also been a contributing factor.

Table 5.3   An Example of Event-resistance in One Area Resulting from Two Closely-spaced Events with Similar Triggering Rainfalls, Wellington, New Zealand (Eyles et al. 1978)

Date	Maximum rainfall (mm) 24 h	72 h	Rainfall in preceding 4 months (mm)	Landslides Number	Volume (m^3)
7-10 October 1974	61	109	656	46	161
15-17 July 1976	60	115	413	9	24

If triggering events on certain types of terrain are separated by long intervals, there may be sufficient time for slope-ripening to occur. This process will lower the threshold by an amount related to the time elapsed since the last event. On the other hand, some events have been observed to produce an immediate weakening of the slope. This condition can be produced in rapidly draining terrain when rainfall stops and landslides are left in a partially evacuated state. The consequent loss of peak strength through shearing and the formation of localised catchment areas by slope deformation can lower the threshold value required for future movement. Apart from the wide variation in thresholds from place to place, triggering thresholds can thus also be seen to vary in time.

The predictive capability of a climatic parameter also relates to how well it represents the amount of water within the regolith. Day-of-event rainfall provides only an approximation of critical soil water status during a landslide event. Soil water status (SWS) for a unit of time at one site can be expressed as:

$$SWS = SWS_a + (P-E) + (Di-Do) \qquad (5.1)$$

where SWS_a is the antecedent soil water status at the
beginning of an event,
P is event precipitation which infiltrates into
the regolith,
E is evapotranspiration during the event,
Di is incoming subsurface drainage, and
Do is outgoing drainage.

5.2 TRIGGERING THRESHOLDS

Eyles et al. (1978) have established approximate rainfall-
related, landslide-triggering thresholds for the steep
greywacke terrain of Wellington City, New Zealand (**Table 5.4**).
These have been determined principally by analysing the climatic record and correlating this with the history of landslide
episodes (**Figure 5.5**; **Tables 5.5 and 5.6**). Spatial correlation of the limit of landslide occurrence and storm isohyets
(**Figure 5.6**) for a large landslide episode that occurred in
summer (**Figures 5.7 and 5.8**) has also provided estimates of
landslide thresholds for Wellington City. For example, cut-
and-fill slopes within the city produced significant failure

Table 5.4   Landslide-triggering Rainfall Thresholds (mm)
for Major Failure of Cut-and-Fill Slopes,
Wellington City, New Zealand (after Eyles et al.
1978)

	Year	Four monthly	Monthly	72 h	24 h
PTn	1370	750	250		
PTx	1800	800	-		
PTx wet antecedent conditions				109	
PTx dry antecedent conditions				120 (150)	

Note: PTn is the minimum probability threshold and
PTx is the maximum probability threshold.
The figure in brackets is determined by spatial
correlation, whereas the other data have been determined by analysis of the 100 year rainfall record.

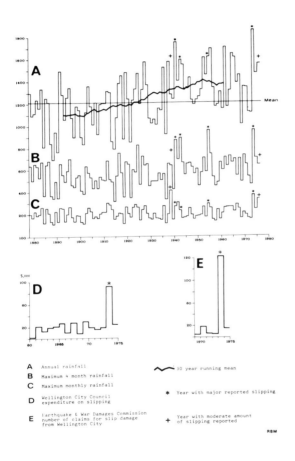

A Annual rainfall
B Maximum 4 month rainfall
C Maximum monthly rainfall
D Wellington City Council
  expenditure on slipping
E Earthquake & War Damages Commission
  number of claims for slip damage
  from Wellington City

∿ 30 year running mean

* Year with major reported slipping

+ Year with moderate amount
  of slipping reported

RBM

Figure 5.5   Indices of landslide activity and rainfall
              through time, Wellington City, New Zealand
              (Eyles et al. 1978)

Table 5.5   Storms at the End of Dry Periods which have Caused Significant Landsliding in Wellington City, New Zealand (Eyles et al. 1978)

Date		Duration (hours)	Rainfall (mm)	Station
26 February	1911	24	161	Karori
2 May	1913	24	145	Brooklyn
20 December	1924	24	123	Thorndon
3 February	1936	48	122	Kelburn
12 December	1939	24	169	Karori Reservoir
28 December	1939	24	156	Beacon Hill
20 December	1976	24	153	Kelburn
20 December	1976	24	245	Karori

Source: 1911 to 1939 Evening Post; 1976 field survey and Meteorological Service records.

Table 5.6   Landslide-producing Storms 1974, Kelburn, Wellington City, New Zealand (Eyles et al. 1978)

Date	Rainfall (mm)		72 h Return
	24 h	72 h	Period (yrs)
14-16 April	80	116	3.5
27-29 May	83	118	3.6
2-4 July	88	124	4.1
7-9 October	60	109	2.9

when rainfall amounted to 150 mm in 24 hours while unmodified slopes required between 200 mm and 250 mm of rain in 24 hours before failing.

For the Wollongong district of Australia, Young (1978) has used monthly rainfall data to establish landslide thresholds (Table 5.7). Although a PTx of 750 mm monthly rainfall can be identified, a wide margin of probability is evident, with landslides occurring on a few occasions even when monthly rainfall is less than 125 mm. Correlation of day-of-event, 24-hour rainfall with landslide events may also produce a wide margin of probability. For example, landslide-triggering 24-hour rainfalls on the Otago Peninsula, New Zealand, during 1977 and 1978, yielded PTn = 6 mm and PTx = 57 mm. The PTn was exceeded on 79 days without producing landslides (Crozier 1982a).

179

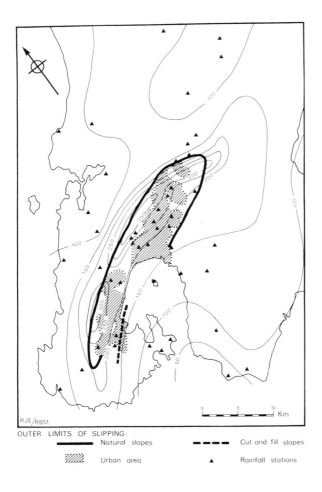

OUTER LIMITS OF SLIPPING:

    ━━━━━   Natural slopes   ━ ━ ━ ━   Cut and fill slopes

    ▨▨▨   Urban area   ▲   Rainfall stations

Figure 5.6   Distributional limits for landslides on natural and cut-and-fill slopes in relation to 24-h rainfall depth (mm) from 19-20 December 1976 Wellington/Hutt Valley, New Zealand (Eyles et al. 1978; rainfall, Tomlinson 1977)

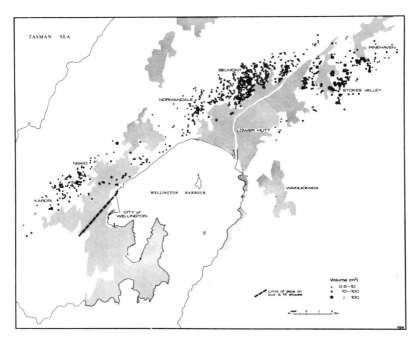

Figure 5.7   Landslides on natural slopes, 20 December 1976
episode, Wellington/Hutt Valley, New Zealand
(Eyles et al. 1978)

5.2.1   EXCESS PRECIPITATION

Attempts to refine climatic triggering thresholds have
included some of the factors represented in equation 5.1. For
example, Young (1978) attributed an increase in the incidence
of mass movement activity with altitude to an increase in
excess precipitation (precipitation minus evapotranspiration).
Using the same approach, Eyles (1979) was able to narrow the
range of thresholds for 24-hour triggering rainfalls for
landslide activity occurring between 1930 and 1977 in
Wellington City. He concluded that excess rainfall above
50 mm to 55 mm results in 'slipping', 60 mm to 90 mm excess
rainfall resulted in the 'possibility' of major slipping' and
100 mm or more excess rainfall resulted in 'major slipping'.
'Slipping' events had return periods of 1.9 years and major
'slipping' events had return periods of 3.4 years.

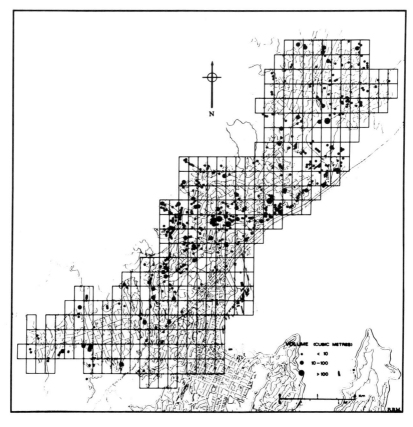

Figure 5.8   Landslides on modified slopes, 20 December 1976
episode, for part of Wellington City, New
Zealand (Eyles et al. 1978)

5.2.2   RAINFALL INTENSITY

The importance of rainfall intensity, often in conjunction
with other climatic parameters, has been noted by a number of
authors. From porewater pressure measurements in the field,
Pierson (1977) found that the variation of piezometric head
was most closely associated with 24-hour rainfall and that
debris slides occurred with rainfalls between 130 mm and
140 mm which included six-hour bursts of at least 50 mm to
75 mm. Using controlled experiments, Okimura (1983) attri-

Table 5.7   Relationship of Monthly Rainfall to Occurrence
of Landslide Events 1892-1974, Wollongong,
Australia (Young 1978)

Monthly rainfall class (mm)	Number of months in each class	Number of months with slips	Months with slips as % of total months in each class
<125	853	5	0.6
125-250	167	18	10.8
250-375	55	22	40.0
375-500	17	9	52.9
500-625	7	6	85.7
625-750	4	1	25.0
>750	1	1	100.0

buted slope failure to periods of accelerated rise in ground-water. When soil throughflow rates reached a certain level, accelerated rises in groundwater were proportional to throughflow and steady-state soil water movement was achieved. He concluded that under these conditions, the critical acceleration of groundwater rise would be directly related to increases in rainfall intensity.

Caine (1980) has attempted to establish a universal threshold for the triggering of shallow landslides and debris flows. To achieve this he plotted the duration and mean intensity of rainstorms associated with 73 reported landslide incidents (**Figure 5.9**). Although all events are reported to have taken place on natural slopes, they inevitably represent a variety of geologic terrain and have occurred during a range of antecedent moisture conditions. In addition, no probability value can be assigned to the threshold because there is no record of how frequently the threshold is exceeded without producing landslides.

From equation 5.1 it is evident that soil water status at any one site will only increase when the combined rates of rainfall and subsurface inflow (Di) exceed the rate of outflow. Thus if failure occurs during very low intensity rainfall, critical conditions are probably a function of sub-surface lateral water movement and long duration events. Failure under such conditions is also likely to be confined to sites which concentrate subsurface water. There is also an upper limit to the effectiveness of rainfall intensity. This occurs when rainfall intensity exceeds the infiltration rate and overland flow occurs. This limit appears to have been met in the poorly vegetated areas during the Hong Kong landslide episode described by So (1971). The comparatively high incidence of failure within vegetated areas was attributed to greater water access as a result of high infiltration capaci-

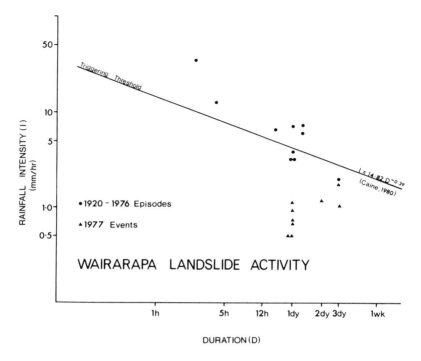

Figure 5.9   Rainfall duration and intensity landslide-
triggering threshold from Caine (1980) with
landslide triggeering rainfall conditions
plotted for episodes from Wairarapa, New Zealand

ties of the vegetated soils compared to those of the poorly
vegetated soils. The upper limit may also vary with seasonal
or antecedent weather conditions owing to their effect on the
infiltration capacity of the soil.   Desiccation cracks developed during periods of dry weather may sufficiently enhance
infiltration rates to lower the triggering threshold.
    Because there are upper and lower limits to the effec-
tiveness of rainfall intensity, mean or maximum rainfall
intensities by themselves are not entirely appropriate
measures of storm conditions.   A more suitable measure, developed by Wieczorek and Sarmiento (1983), is the duration of
rain occurring at or above specific rainfall intensity values.
This is referred to as the 'intensity-duration parameter'
(ID).   For example, ID_6 = 2 hours represents a rainfall intensity equalling or exceeding 6 mm/hour for a duration of two
hours.   In their study of rainstorms in an area near La Honda,

California, between 1975 and 1982, Wieczorek and Sarmiento found that, regardless of the ID value, a minimum value of 11 inches (279 mm) of antecedent seasonal rainfall was required to trigger debris flows. The most significant ID value for distinguishing triggering from non-triggering storms, once the antecedent requirement had been met, was $ID_{0.2}$ inches (5.1 mm/hour) > 3 hours.

5.2.3 ANTECEDENT CONDITIONS

Wet, rather than dry, antecedent conditions appear to be more important in lowering the rainfall-related, landslide-triggering threshold. For example, in the Santa Monica Mountains, southern California, Campbell (1975) noted that a period of rainfall providing at least 6 mm/hour was sufficient to produce debris slides if it were preceded by a rainfall of 250 mm. Corresponding values for the Los Angeles area have been given as 6.35 mm/hour after 255 mm of antecedent rainfall (Nilsen et al. 1976). In Hong Kong, landslide 'disasters' appear to occur when daily rainfall exceeds 100 mm after a 15-day antecedent rainfall in excess of 350 mm. For the six landslide episodes occurring between 1933 and 1969 in the San Dimas Experimental Forest, Rice (1982) assessed triggering conditions as being 500 mm of rain over a period of five days with 150 mm to 200 mm falling in 24 hours late in the storm.

Antecedent rainfall. Because of the widespread availability of climatic data, the Antecedent Precipitation Index (Pa) (Kohler and Linsley 1951) has been used as an index of antecedent soil water content (Crozier and Eyles 1980), and applied to the 10 days preceding landslide event rainfall. The index is given as:

$$Pa_0 = kP_1 + k^2 P_2 + \ldots\ldots k^n P_n \qquad (5.2)$$

where Pa_0 is the antecedent daily rainfall for day 0,
P_1 is the rainfall on the day before day 0,
P_n is the rainfall on the nth day before day 0, and
k is a constant <1.0, in this case 0.84

Figure 5.10 indicates that, in combination with daily rainfall, this index allows the definition of an approximate threshold between landslide-producing conditions and 'stable' conditions. However, PTn and PTx are separated by a wide probability margin (Figure 5.11) and an even more precise threshold needs to be defined for predictive purposes.

Antecedent soil water status model. One successful approach for determining the critical soil water content has been the development of an empirical index termed the Antecedent Soil

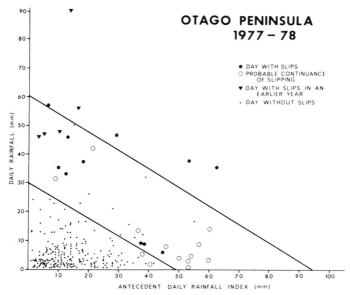

Figure 5.10   Minimum and maximum probability thresholds
derived from daily rainfall and antecedent daily
rainfall index, Otago Peninsula, New Zealand
(Crozier and Eyles 1980)

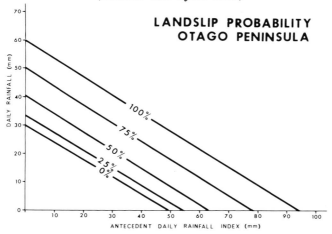

Figure 5.11   Probability of raindays producing landslide
events, Otago Pensinula, New Zealand (Crozier
and Eyles 1980)

Water Status (Crozier 1982a). For conditions below field moisture capacity, this is simply the soil moisture storage as determined by the Penman (1948) water balance equation and for conditions above field moisture capacity, it is a measure of antecedent-excess precipitation EPa over 10 days, defined as follows:

$$EPa_0 = kEP_1 + k^2EP_2 \ldots + k^nEP_n \qquad (5.3)$$

where EP is the daily rainfall in excess of Potential Evapotranspiration and soil storage requirements (soil moisture capacity depends on soil characteristics, and other factors as for equation 5.2)

The success of this approach to defining a threshold between triggering and non-triggering conditions can be seen in **Figures 5.12 and 5.13**.

The strength of the antecedent soil water status model is both its simplicity and its ability to define distinctive thresholds between landslide-producing conditions and stable conditions. It is simple in the sense that the climatic parameters employed are those readily available from standard climatic stations and the necessary calculations are straightforward.

The success of this particular model (as with any historic approach) depends on the establishment of an accurate record of landslide activity. At the very least, the date, time and precise location of landslides need to be recorded on a systematic basis over a period of years. If such records are not available, it is necessary to resort to a careful evaluation of indirect sources such as newspapers, insurance claims or interviews.

The definition of antecedent soil water status in the model requires a number of decisions to be made related to both the hydrological properties of the terrain and the type of landslide involved. Initially, a value for soil moisture capacity must be determined with respect to the porosity of the soil and the depth of the failure plane. In the examples shown, a value of 120 mm was chosen as an average of the water-holding capacity of the regional soil types within the top 76 cm of the regolith.

Another assumption incorporated in the model is that rain water in excess of soil moisture requirements will be a factor in determining stability conditions. Although over a period of time, excess water may be lost from a particular site of input by drainage and evapotranspiration, much of it will slowly migrate to areas of moisture concentration within the slope. It is these temporary storage areas, where subsurface flow is concentrated or inhibited, that generally give rise to landslides.

The rate of water loss from the slope is approximated in

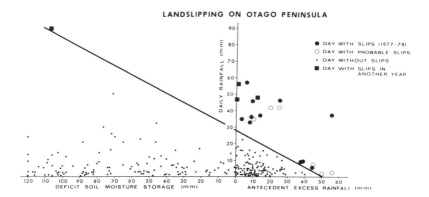

Figure 5.12   **Triggering thresholds derived from daily rainfall and antecedent soil water status (storage and antecedent precipitation index applied to excess rainfall), Otago Peninsula, New Zealand** (Crozier and Eyles 1980)

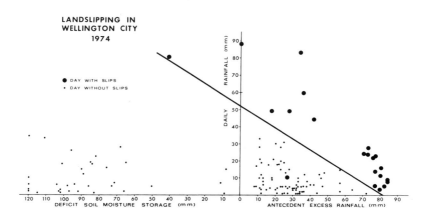

Figure 5.13   **Triggering thresholds derived from daily rainfall and antecedent soil water status, Wellington City, New Zealand** (Crozier and Eyles 1980)

the model by the constant k but this will vary not only with evapotranspiratory controls but also with soil type and slope angle. A value of 0.84 is used in the examples shown as it was found by trial and error to be successful in delineating a distinct threshold. The k factor is decayed by an exponential function so that past rainfall exerts progressively less influence on the soil water status index as time elapses. The equation expressed in this form allows for greater absolute outflows with higher soil water levels, in recognition of the lower soil tensions and greater hydrostatic heads that prevail under such circumstances.

Variation in micro relief and soil conditions means that the soil moisture status determined by these indirect methods can only be considered to be a broad regional index of actual conditions.

The model could be further improved by using storm rainfall as opposed to daily rainfall and increased precision would result from employing a more precise measure of rainfall intensity. However, increased refinements require additional data and calculation time which may inhibit the effective usefulness of the method for some purposes.

5.3   APPLICATION OF THRESHOLDS

Triggering thresholds could theoretically be determined by correlating direct groundwater observations with the initiation of landslides. However, there are few data available for this purpose and the climatic record usually provides a more comprehensive data-base for establishing triggering thresholds. Because of the existence of established climate recording networks, climatically-determined thresholds can also be used to monitor the onset of critical conditions over a wide area. For example, the Antecedent Soil Water Status Model (see Section 5.2.3 ANTECEDENT CONDITIONS) can be used in the following way to warn of the onset of triggering conditions.

The first requirement for the model is the establishment of a threshold line, as described in Section 5.2.3, for the region under study. The antecedent soil moisture assessment is then kept up to date on a daily basis by an easily-made calculation:

When field capacity is exceeded;
$$EPa_0 = k(EPa_1 + EP_1) \qquad (5.4)$$

where factors are as for equation (5.3) and

when soil moisture is less than field capacity,
$$DS_0 = DS_1 - (P_0 - PE_0) \qquad (5.5)$$

where DS_0 is deficit soil moisture storage on day 0,

DS_1 is deficit soil moisture storage for the day before day 0,

P is the precipitation on day 0, and

PE_0 is the potential evapotranspiration on day 0.

For any given day, the soil moisture status can be applied to the threshold graph to obtain the magnitude of daily rainfall required to produce landslides. If a frequency-magnitude analysis of the historical rainfall is carried out then the probability of any landslide-triggering rainfall value can be determined (**Figure 5.14**). For monitoring purposes, the daily soil water status index can be applied to **Figure 5.15** to obtain the probability of occurrence for a landslide-triggering rainfall. If the thresholds have been clearly defined this value can be taken to indicate the probability of landslides being triggered on the following rain-day. Similarly, on a given day, the depth of rainfall can be continuously monitored as it approaches the threshold value for landslide-triggering rainfall, as indicated by the soil water status index. Depending on the vulnerablity of the region and past experience of landslide damage, suitable warning and mitigating measures can be taken by the appropriate authorities.

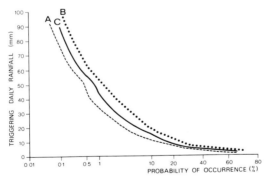

A OTAGO PENINSULA 1918–80   B WELLINGTON CITY 1928–70   C WAIRARAPA 1940–74

Figure 5.14   **Probability of 24-h rainfall equalling or exceeding given values for New Zealand localities** (Crozier 1982a)

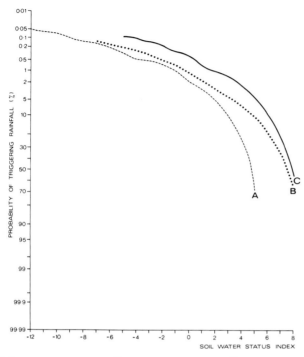

A OTAGO PENINSULA 1918-80   B WELLINGTON CITY 1928-70   C WAIRARAPA 1940-74

Figure 5.15   Probability of a 24-h rainfall triggering
landslides at given values of the soil water
status index for selected New Zealand localities
(Crozier 1982a)

A second major application of the antecedent soil water
status model is its use in classifying the sensitivity of different terrain types to landslides brought about by rainfall.
When threshold lines are superimposed on the same graph, the
sensitivity of the regions can be ranked by comparing the
relative positions of the threshold lines.   In **Figure 5.16**,
A is more sensitive than B which in turn is more sensitive
than C.   These lines can be graphed in terms of the probability of occurrence to indicate the susceptibility or likelihood of occurrence of landslides within each region.

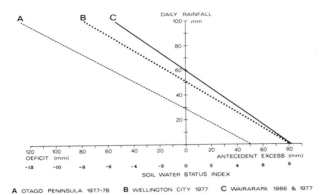

A OTAGO PENINSULA 1977-78   B WELLINGTON CITY 1977   C WAIRARAPA 1966 & 1977

Figure 5.16   **Regional sensitivity to landslide – triggering rainfalls for selected New Zealand localities**

LIVING WITH LANDSLIDES

6

Landslides, like many other high-energy, natural processes, only become hazards when people, property and livelihood become threatened. The problem is that over the last 100 years there has been an unprecedented increase in the number of people and the amount and value of property at risk. The area required to provide food and other primary resources has increased as a result of the human demand for such resources. The relationship between these increases and the chance of landslide damage is not simply arithmetic. This is because pressure on the resources has forced people to exploit hazardous areas which are only marginally suited for the uses to which they are put.

Another factor which increases the vulnerability of society to landslide damage is urbanisation, because it concentrates people and fixed assets in a limited area as well as increasing reliance on transport, communication, and service systems. Of similar importance has been the development of mankind's technical capability to modify the physical environment and the extent of the changes that have therefore taken place. In many instances, landscape modification has led to increased sensitivity of the land surface and an increase in the frequency of landslide activity (see for example, Section 4.5.3 VEGETATION CHANGES).

The vulnerability of society to landslides can also be increased as a result of economic and political pressures which, in some countries, have forced people into environments, land use practices and socio-economic systems with which they are unfamiliar (Susman et al. 1983). People in these circumstances may have insufficient knowledge of the endemic hazards, inappropriate adjustment mechanisms or too few resources for mitigating these hazards. Poor countries in particular have suffered greatly as a result of many of these recent trends, particularly in terms of loss of life (Crozier 1982b). Developed countries however, have not escaped the costs of landslide activity and other natural hazards; instead the costs have been transferred - loss of life from most natural hazards has decreased but economic costs have increased dramatically (White and Haas 1975).

Communities in transition between subsistence and cash-based, developed economies, can be particularly vulnerable. For example, as a result of cyclone Wally which occurred in Fiji in Easter 1980, 21 villages suffered severe damage and 2000 families qualified for emergency subsistence rations (Crozier et al. 1981). Lack of food was the immediate problem, as a result of destruction of gardens and livestock by landslides and flooding and the area's semi-dependence on a cash-based economy. For example, in one village, 17 families had been persuaded to take up cocoa growing as part of an external aid project. The villagers had acquired a cocoa drying plant and consequently their garden food crops had largely been replaced by the cash crop. During the cyclone, the drying plant was buried by a landslide, access to markets was inhibited by cyclone damage and the families were forced to become reliant on relief supplies to survive. In this situation the inherent resilience of the traditional subsistence system had been lost without gaining the benefits of a more developed society.

One almost accidental outcome of scientific research on landslides is the lessening of their status as 'acts-of-God'. The acquisition of knowledge and its communication endows a certain level of responsibility. For instance, under some jurisdictional systems, communication of knowledge is tantamount to advice and advice is treated as a commodity. Legal action for negligence may be taken if that advice is proved misleading or inadequate. The replacement of fatalism by a sense of responsibility has also followed a heightened awareness of the problem brought about by improving news media coverage.

Society is still struggling with just how this responsibility should be handled. Who should fund research on the landslide problem? What sector of the community should be responsible for taking precautions against landslide damage? Should government simply acquire relevant information, communicate it and let the land users decide to what extent they will take precautions or should the government go further and determine the nature of those precautions? When landslide damage occurs, who should bear the costs? Within western political systems, the more control exerted by government, the more responsibility they are seen to hold and the more they are looked to for compensation. Most of these political systems, however, either in the name of political freedom or economic necessity, acknowledge only partial responsibility, leaving the rest to be borne by professional groups or individuals. With government accepting only partial responsibility, the legal system or ad hoc commissions of inquiry are frequently left to identify responsibility and award compensation.

In many less developed countries, individuals have little
recourse to legal action and little hope of obtaining compensation. If they suffer enough or are in politically strategic
locations, they may attract relief aid from international
sources. Thus, the extent to which responsibility is realised
and acted upon depends on the perceived severity of the risk,
the socio-economic and scientific level of development and the
political philosophy of the country concerned.

6.1 LANDSLIDE-RELATED COSTS

The costs of landslide activity can be classed as:
1. personal: death, injury, and prolonged psychological
   and physical health problems;
2. economic loss: individual and public; and
3. environmental damage.

Some of these costs are immediate and can be related
directly to the landslide event; others are manifest over long
periods or are included with other general planning and management activities and may be related only indirectly to
landslide activity. Many landslide episodes are associated
with other simultaneously occurring hazards, a fact which also
makes it difficult to isolate costs relating specifically to
landslides. Landslide-triggering rainstorms, for example,
often produce damaging floods as well as landslides and,
during earthquakes, damage is caused by collapsing structures
as well as by slope movements.
Obtaining data of sufficient accuracy or representative-
ness to illustrate the real costs and trends of landslide
hazards is extremely difficult if not impossible to do at present. International organisations including the United
Nations Disaster Relief Organisation have been keeping
detailed records of events only since the early 1970's.
However, some individual countries, such as the United States
and Japan, have comprehensive records for longer periods but
these are exceptions.

6.1.1 PERSONAL COSTS

Death by landslide is a tragically obvious personal loss. It
may also involve the trauma of economic, social and emotional
readjustment on the part of the bereaved family or community.
Such effects combined with trauma resulting from proximity to
the event itself may require a costly process of adjustment.
A sometimes neglected aspect of large-scale disaster is the
debilitating psychological effect experienced not only by
those directly affected by the event but also by some relief
workers (Taylor 1983). Injury also has long-term costs as it
may not only reduce the social and productive role of an indi-

195

vidual but may also impose added costs for medical treatment and support.

6.1.2 ECONOMIC COSTS

The direct economic consequences of a given magnitude of landslide event depend on the nature of society and the type of landscape affected. The greatest immediate costs usually result from impact on transport and communication systems and buildings (**Figures 6.1 and 6.2**). For example, mudflows that accompanied the catastrophic eruption of Mt St Helens, Washington State, United States, in May 1980 '...damaged or destroyed more than 200 buildings, ruined 44 bridges, buried 27 km of railway and more than 200 km of roadway, badly damaged three logging camps, disabled several community water supply and sewage disposal systems and partly filled channels and reservoirs' (Schuster 1982). However, some events are not selective. The now-hardened landslide debris which covers the 18 000 inhabitants of Yungay, Peru - their church, their houses, their town square, their troubles and their hopes - should be a constant reminder of the indiscriminate potential power of landslides.

Small-scale landsliding, on the other hand, involves lesser but more frequent direct costs. These may include debris clearance from streets, drains, streams and reservoirs; new or renewed support for road and rail embankments and slopes; minor vehicle and building damage; personal injury; livestock, timber, crop and fencing losses and; damaged utility systems (**Figure 6.3**).

The indirect costs related to landslides can be divided into preventative costs and event-related costs. Event-related indirect costs include:

1. mobilisation and support of relief and civil defence personnel;
2. temporary or replacement housing;
3. supplying food and financial support for victims;
4. compensation for other personal or community costs;
5. costs of delays to transport or costs of alternate routing;
6. losses of human, land, and industrial productivity especially degradation of soil and timber resources;
7. costs of providing substitute grazing or stock feed and raw materials for industry;
8. opportunity loss on damaged land or destroyed resources;
9. depreciated land values;
10. land rehabiliation; and
11. costs of legal action or commissions of inquiry into causes, responsibility and culpability.

Preventative costs can be even more difficult to assess and include:

Figure 6.1   Landslide in the base material of a main north
road near Dunedin City, New Zealand

1. costs of research into many aspects of the landslide
    problem including field surveys, mapping, laboratory
    tests and analysis;
2. establishment, management and inspection of damage
    reduction policies and regulations;
3. design costs on earthworks;
4. opportunity losses on regulating or preventing cer-
    tain land use activity on unstable land;
5. additional costs relating to the transfer of activi-
    ties to stable land in more distant locations;
6. costs of implementing preventative or control
    measures including planting, drainage and other
    treatment;

Figure 6.2   Section of the Graben at the head of the
              Abbotsford blockslide which occurred near
              Dunedin City, New Zealand in 1979. Sixty-nine
              houses were destroyed (Photo: Otago Daily Times)

Figure 6.3   Vehicle damage as a consequence of cut-slope failure, Wellington City, New Zealand

Figure 6.4   Habitat alteration on slopes and in stream chan-
nel as a result of debris slides, Katikati, New
Zealand (Photo: Ministry of Works and
Development Photo Library, D Bircham)

Figure 6.5   Habitat alteration and drainage interference
resulting from large slump, Kairakau, southern
Hawke's Bay, New Zealand (Photo: P R Stephens)

7. maintenance of preventative or control measures;
8. implementation and maintenance of warning systems;
9. costs of issuing 'false' warnings; and
10. costs of communication and education.

6.1.3 ENVIRONMENTAL COSTS

Unless a landslide event is very extensive or the morphogenic system within which it occurs is very unstable, its environmental impact is usually short-lived. However, in inherently unstable areas, which provide frequent or large volume slope movements, both the slope and downstream consequences may be severe and long-lasting (**Figures 6.4 and 6.5**). The principal impact on the slope is the change in the amount of in-situ, biologically and pedologically 'developed' habitat compared to the amount of immature substrate, consisting of erosion surface and debris. These changes have been conventionally viewed as a form of habitat degradation. The removal of soil and its biomass is seen as interrupting the long-term trend towards a mature ('climax') community. However, just as fire may operate as a natural agent necessary for the continued survival of a healthy, multi-specied plant community, a similar role may be played by landslides. Moss and Rosenfeld (1978) suggest that instead of landslide activity being viewed as an agent of degradation, in some areas, it may represent an important element in the maintenance of species diversification and re-vitalisation of habitat. A number of studies have examined form and rate of the recolonisation and soil development of landslide surfaces (Langenheim 1956; Flaccus 1959; Mark et al. 1964; James 1973).
The disruption of the land surface as a result of slope movement may also derange the drainage system, altering drainage pathways and creating poorly drained depressions and ponds.
The effect of landslide debris which reaches the stream channel depends on its rate of input, calibre of sediment and its stability with respect to fluvial activity. Some mass movements in this situation may develop into fast moving debris torrents which can completely strip riparian vegetation, surface soils, and channel sediments from many kilometres of channel, during one event (Grant et al. in press). In the process, micro fauna and flora are depleted and consequently fish habitats are destroyed. Chronic sediment input from eroding landslide scars and continued disruption of bedload can also damage fish habitats (Phillips 1971; Meehan 1974). High suspended sediment concentrations can clog spawning gravels and prevent embryo development, destroy food sources and fatally damage the gill structure and respiratory systems of fish.
Relatively stable landslide deposits within the stream channel may bring about bank erosion in downstream areas and

accelerate the development of meanders, prevent fish migration and form debris dams which could eventually collapse with catastrophic consequences. In other situations, 'stable' landslide debris within the channel may encourage a biologically-resilient diversification in an otherwise uniform stream habitat.

6.2 SOME COST ESTIMATES

The scientific literature records only a few of the damaging landslide events which take place. **Appendix One** lists 38 events from around the world which are occasionally referred to and these alone count for at least 149 000 fatalities. Even from a limited list such as this, the importance of rainstorms and earthquakes as landslide-triggering factors and the high representation of Pacific-rim countries is clearly evident. In Japan, for example, during the six years between 1967 and 1972, 662 people were killed by debris flows and 682 by landslides, together accounting for 57% of all fatalities brought about by natural disasters within the country during this period (Ikeya 1976).

The extent of the problem in Peru, another Pacific-rim country, has been documented by Taype (1979). He showed that, in a period of 50 years, there have been 168 important slides, 37 mudflows and 3734 huaycos (debris torrents), accounting for 34 975 deaths and losses amounting to US$26 million. In the light of regional studies such as this, records kept by some international agencies need to be interpreted carefully. In **Table 6.1** for example, only 6065 deaths have been attributed to landslides and avalanches from around the world in the period between 1900 and 1976. These figures exclude numerous small but fatal events and some landslide fatalities are probably included within the other natural hazard categories.

Economic costs of landslide events have contributed significantly to the overall impact of natural hazards on the economy of many countries. Damage caused by natural disasters in the five countries of the Central American Common Market is estimated to have reduced the annual growth rate of 'gross domestic product' by approximately 2.3 percent over the 15 year period from 1960 to 1974 (United Nations 1979). In another example, it has been estimated (Sidle et al. 1985) from mass movement erosion rates and their long-term effect on pasture productivity, that mass movement erosion has the potential to reduce the export revenue of New Zealand by 5%, or several hundred million dollars, over the next few decades. This could have serious consequences for the country which is a small trading nation relying heavily on an agriculture and forest based economy.

Although the United States economy is more resilient than that of New Zealand it still has to cope with direct and

**Table 6.1   Results of Natural Disasters Between 1900 and
1976** (United Nations 1979)

	Persons killed	Persons injured or left homeless
Earthquakes	2 662 165	28 894 657
Volcanic eruptions	128 058	337 931
Floods	1 287 645	175 220 220
Landslides	3 006	44 673
Avalanches	3 059	150
Cyclones	434 894	17 648 463
Hurricanes	18 513	1 197 535
Typhoons	34 103	5 437 054
Storms	7 110	3 432 641
Tornadoes	1 175	342 459
Total	4 579 728	232 555 783

Source: International Red Cross

indirect costs of landslide activity amounting to over $1 billion annually (Schuster 1978). Petak and Atkisson (1982) estimated that in 1979, building losses alone amounted to $370 million and, assuming no change in building standards, annual losses could reach $871 million by the year 2000. However, if appropriate mitigating measures are introduced this figure could be reduced by 60%. By their estimate, potential landslide damage is more amenable to cost reduction than seven out of the eight other natural hazards studied.

Most house and building owners in New Zealand are covered for landslide damage through a compulsory premium levied on their fire insurance. The record of claims against this insurance provides some measure of direct costs resulting from landslide damage within the country (**Table 6.2**). In the 11 year period between 1971 and 1981, 2388 claims involved a total amount of over NZ$4 million (Hellberg 1984). More comprehensive indications of landslide costs in urban areas have been obtained for the San Francisco Bay region. For the winter of 1968 to 1969, Fleming and Taylor (1980) estimated total costs from landslide damage of between $33 million and $56 million and during the 1972 to 1973 rainy season, total costs amounted to $9.6 million (Nilsen et al. 1976).

Even in less densely populated areas, large single events may still prove expensive. In the space of about five minutes, the clay landslide of 30 November 1977 at Tuve, southwestern Sweden, destroyed 65 single-family houses, 27 hectares of farmland and killed nine people. This one event is estimated to have cost the community US$30 million. In this

Table 6.2   Earthquake and War Damage Act, New Zealand,
             Landslip Damage Claim Statistics

Year	Number of claims	Amount ($NZ)
1971	181	109 561
1972	140	65 884
1973	113	41 371
1974	90	35 093
1975	322	226 972
1976	270	194 130
1977	284	432 824
1978	217	346 367
1979	150	186 735
1980	460	2 350 405
1981	161	116 808
Total	2388	4 106 150

Source: Appendix A, Hellberg (1984)

region, Cato (1982) notes that landslides greater than one
hectare in area occur on average every two years. Within
historic times these have accounted for direct and indirect
costs averaging US$10 million to US$20 million per year.
An example of economic loss to pasture production caused
by one episode has been given by Hawley (1980). In the winter
of 1977 a landslide episode affecting about 1400 km^2 of
Wairarapa hill country, New Zealand, caused immediate loss of
production totalling NZ$0.6 million. Despite standard over-
sowing and fertiliser application, the cumulative loss of production over the succeeding 14 years is estimated (in constant
dollars) at NZ$5.6 million and over 55 years at NZ$9.35
million.

6.3   MITIGATING COSTS

Assuming human responsibility is accepted for the landslide
problem, what is required to mitigate successfully landslide-
related costs? Looking at the problem as both an individual
and community concern, seven requirements can be identified:
    1. a technical information base;
    2. an informed populace;
    3. an informed and capable local government;
    4. a philosophical basis for distribution of costs;
    5. an appropriate statutory and legal infrastructure;
    6. an informed and capable professional and technical

community for managing and executing a mitigation system;
7. an effective system of communication and education.

The direct options for mitigating the landslide hazard include:
1. permanent avoidance of unstable areas - usually achieved by government land acquisition or, in some cultures, by traditional tribal decrees such as New Zealand Maori 'Tapu';
2. temporary avoidance of unstable areas; for example, evacuation with the onset of critical conditions involving the use of monitoring and warning systems and seasonal use of hazardous areas;
3. restricted land use activities or controlled density of use; for example, use restricted to low value, low vulnerability, or non-aggravating activities;
4. imposition of building design standards;
5. prevention of potential landslides by engineering works or other treatments;
6. control of landslide activity; for example, installation of rock-fall chutes, debris run-out areas, or tree planting to reduce movement rates; and
7. relief, compensation, repair and replacement; for example, by insurance, government grants or charity.

The degree to which any one or more of these options is adopted depends to a large extent on the political philosophy and level of socio-economic development of the community. Undoubtedly, most of these measures could not be implemented adequately without government land acquisition and regulation (United Nations 1980). Although these measures provide freedom from some of the consequences of slope instability, they take away other freedoms; this is a fundamental philosophical dilemma, the solution to which varies from country to country and from time to time. Whatever mitigating measures are exercised, they are costly to implement. The question of how those costs should be shared between the affected individual and the community at large represents the other major philosophical dilemma in coping with the landslide hazard.

In the most extreme case, a government may decide to accept total responsibility for landslide damage or for certain forms of damage. For example, all personal injury from landslide or indeed from any other cause in New Zealand is compensated for through the Accident Compensation Act of 1956 (New Zealand Law Society 1984). All-embracing coverage of this sort removes the necessity for court action and reliance on common law which otherwise would be the normal means for determining fault, culpability and compensation. In effect, it spreads the cost burden throughout the community by way of indirect taxation. At the same time, assumed responsibility places on the government the onus of research, regulation and the implementation of preventative measures.

In New Zealand, another type of cost spreading is used to compensate for landslide damage to property, in the form of the Earthquake and War Damages Act of 1944. Initially designed to cover war and earthquake damage, this insurance scheme was extended to cover landslip damage on a voluntary basis in 1956 and on an automatic basis in 1970 (Hellberg 1984). Cover is financed by a compulsory levy on all fire insurance premiums and, as such, covers most property owners against building damage. An argument sometimes raised against this form of insurance is that those who have been wise enough to purchase land free from hazard or who otherwise take their own precautions are unlikely to receive any benefits from the scheme. In other words, the cautious may be seen as sub-sidising the folly of others. This dilemma is partly addressed under the insurance scheme, as land itself is not covered. Perhaps a fairer method of financing this form of insurance cover is 'premium loading', based on assessed potential hazard. This, however, is complex to administer and assumes an adequate knowledge of hazard occurrence as well as involving the cost of scientific investigation required to establish a hazard zoning system.

In the absence of a pre-determined and guaranteed system of compensation, relief and compensation may be met by informal and ad hoc arrangements. From time to time governments or international agencies may grant special relief funds, offer low interest, re-construction loans, or supply relief resources. These measures offer a precarious form of security which generally operates only for large-scale disasters and to an extent determined by the prevailing economic and political conditions.

A common way of mitigating potential damage is by statute and regulation. In some jurisdictional situations, local government or national agencies are required by statute to 'take into account', 'reduce', or 'prevent' damage by landslide; in others, the power may exist to refuse permission to build in areas subject to landslide hazard. In addition, local government is usually empowered to pass by-laws in the interests of the community they serve.

In pursuing their responsibilities, regulating bodies can employ a range of mitigating measures including building codes, earthwork standards, and land use controls. The eagerness with which existing powers are used or a particular option is adopted depends not only on the margin of benefit over cost for any proposed measure but also on the way in which the hazard is perceived. The extent to which the problem is appreciated often depends on how well scientific information is communicated. Paradoxically, the most eloquent advocate is the event itself; the most effective mitigating measures are often implemented within a short period of a

serious event. There is an understandable tendency to rely on technological and engineering measures for protection against
landslides. With this approach the end-user generally pays
indirectly for the required work, but the option of use is
preserved. Prohibiting occupance of land subject to hazard,
on the other hand, may appear to be wasting a resource. The
engineering solution may also be attractive to local authorities because of its potential to provide employment and subsequently tax revenue. Technological measures, however, do
not always provide sufficient protection and in the event of
failure their destruction may add to the cost of landslide
damage.

Many schemes employed for reducing landslide damage are
still too new to allow an evaluation of their performance.
However, encouraging results have been obtained in the City of
Los Angeles since the enactment of a grading code in 1963.
Damage to private property subject to this code has been
reduced by more than 90% compared with private property
constructed before 1963 (United States Geological Survey
1982). On the basis of these results it is estimated (Alfors
et al. 1973) that damage amounting to nearly $9 billion could
be prevented in California over the period 1970 to 2000 at a
benefit-cost ratio of 9:1.

6.4 DETERMINING LANDSLIDE HAZARD AND RISK

The technical information base for the design of mitigating
measures should ultimately give rise to a statement on
landslide hazard and risk. Fortunately both the activity of
external stability factors and the influence of inherent factors vary in a more or less predictable manner from place to
place. For this reason it is possible to identify zones and
construct maps showing different degrees of landslide hazard
and risk. Ideally, the procedure involves five tasks:
1. identification of the nature, degree of activity and
   critical levels of external destabilising factors;
2. identification of the physical response of inherent
   factors to the critical levels of activity of the
   external factors; that is, a determination of terrain
   sensitivity;
3. integration of both the frequency of occurrence of
   critical levels of the external factors and terrain
   sensitivity to produce a measure of the probability
   of landslide occurrence;
4. combination of the probability of landslide
   occurrence with mass movement characteristics, such
   as rate, depth, volume, and zone of influence to produce an assessment of potential landslide hazard. In
   effect, this is a statement of the frequency/

magnitude characteristics of the phenonemon;
5. combination of potential landslide hazard with the potential human, economic and environmental damage to produce a statement on landslide hazard risk.

In practice, however, there are often insufficient data to carry out all five tasks. If only a small area is under consideration, task 5 (evaluation of the risk) is likely to be considered early in the assessment. If potential risk is considered to be high, then both surface and subsurface investigations may be carried out. A full stability analysis would ensue, combining tasks 1 and 2 to achieve a statement on probability of occurrence. Knowledge of the form, material, volume, and site characteristics, combined with any precedence assessment, would form the basis for a statement of the potential hazard.

On a regional scale, time and financial constraints usually prevent special subsurface testing for all areas under consideration; instead use has to be made of existing information or data which can be rapidly obtained.

The first decision to be made in a regional survey is the choice of the primary 'information unit' or mapping unit. The size and nature of this unit depends on the scale of the area under consideration, the time and financial resources available, the degree of detail of the existing information and the degree of terrain homogeneity. If little pre-existing information is available, the information unit may be arbitrarily defined or based on cadastral or jurisdictional boundaries. Subsequent mapping units, subdivisions, and hazard zones are then constructed solely on the basis of the acquired data. Generally, however, existing geological, soil, vegetation, slope, or geomorphic units serve as the primary mapping units. Brabb (1982), for example, in preparing a map depicting landslide susceptibility for San Mateo County, California, used geological units as the primary map unit and sub-divided these on the basis of slope to form a geology/slope unit. This unit was then reclassified with respect to past landslide behaviour to provide the ultimate units of landslide susceptibility. The method is simple, rapid, and relatively successful in providing a regional assessment of the landslide hazard at a level suitable for basic planning decisions.

Having established the primary mapping unit, each one may be categorised on the basis of its current exposure to external destabilising factors. This can include such factors as: presence and frequency of coastal or stream undercutting, efficiency of drainage or stormwater systems, and, on a broader scale, frequency and magnitude of triggering factors, such as certain weather conditions, seismic activity or particular land use practices. Ideally, analysis of the type discussed in Chapter Five, CLIMATIC TRIGGERING OF LANDSLIDE

EPISODES, should be carried out. This would allow each unit to be assigned a value representing the minimum return period of any potentially destabilising activity.

Task 2, however, conventionally constitutes the principal basis for hazard assessment. Inherent stability factors are surveyed, numerically scaled, and recorded on the primary mapping unit to provide an assessment of terrain sensitivity. An example of using a semi-quantitative scaling procedure of inherent factors to produce an overall measure of terrain or slope sensitivity, is shown by Selby's (1980) rock mass classification (for the parameters used see Section 3.3.11 MATERIAL AND STRUCTURE). In general, however, as many factors should be recorded as the constraints of the exercise allow. **Appendix Two** lists some of the factors which may be ranked in classes, on a scale from 'potentially stable' to 'potentially unstable'. Numerical rankings of individual factors may be weighted and summed to provide aggregate scores for each mapping unit which, in turn, can be ranked to provide an overall assessment of terrain sensitivity.

In task 3, scores representing the incidence of triggering conditions (task 1) are combined with those representing inherent stability conditions (task 2) to provide a measure of probability of occurrence. Thus high inherent stability combined with low minimum return periods for the activity of external factors indicate high probability of landslide occurrence.

It is also useful to record on the map the factors which are most influential in determining the overall probability of occurrence rating. This indicates not only those factors which might be effectively manipulated to improve stability but also those which need to be preserved to maintain stability.

A test of the validity of this method for determining the probability of occurrence can be carried out by correlating the classified units with the occurrence of pre-existing or new landslides.

Tasks 1, 2, and 3 of the above approach arrive at probability of occurrence on the basis of 'pre-determined importance factors'. As such, this method can be employed where there is little evidence of landslide activity. If, however, landslide activity is prevalent in the area, then various forms of locational-factor-analysis and spatial and temporal correlation analysis can be carried out to determine the most influential stability factors within the region. For example, minimum and maximum threshold angles, landslide density/slope angle class, and landslide density or landslide volume can be obtained for each geological unit. This information may be extrapolated to unaffected areas of the same mapping unit class. In the presence of landslides, terrain sensitivity can thus be assessed by 'post-event-determined importance' of stability factors.

If terrain sensitivity can be assessed from existing evidence of landslides, then it may also be possible to classify the units directly into probability of occurrence classes (**Table 6.3**). The criteria used in this process are obtained from the record of landslide activity, the return period of triggering thresholds, morphological and sedimentary evidence and comparison with analogous areas. Crozier (1984) has provided a detailed treatment of the methods and criteria for assessing the frequency and degree of instability from field observations.

In itself, the map of the probability of occurrence (likelihood or susceptibility) does not offer a complete statement of the potential landslide hazard. To define the hazard adequately a prediction of the magnitude of the expected process is also required. A landslide hazard map should therefore indicate the distribution of units subject to different frequency/magnitude levels of landslide activity.

The aspects of magnitude with human and environmental significance include parameters such as: volume, area eroded, rate, depth, degree of disruption, and zone of influence. Inasmuch as landslide types reflect these factors (Crozier 1984), classifying existing or anticipated landslides by type may be a sufficient indication of magnitude. Thus a landslide hazard map may be prepared by classifying probability of occurrence units by landslide type. The actual appreciation of the hazard, however, will depend to a certain extent on the end-use of the information. For example, a soil conservator will be concerned with high probability, shallow soil slides, whereas these may represent only a minor hazard to an urban planner who is concerned with community safety.

Alternatively, magnitude parameters may be assessed and mapped individually or ranked, scored and aggregated to arrive at a magnitude index for each mapping unit. For many types of landslide it may be necessary to extend the basic mapping unit to indicate run-out zones (zones of influence) (**Figure 6.6**). The landslide hazard map developed in this way is the principal scientific contribution to the mitigation of the landslide problem. It is on this information that planners, resource managers or engineers must base their decisions. It is therefore important that the information is clearly, unambiguously, and effectively presented. Lawrence (1981), Varnes (1982), Brabb (1984) and Hansen (1984) have discussed the various methods and techniques available for preparing, presenting, interpreting, and using such maps.

One immediate use for the landslide hazard map is as a basis for assessing risk. Superimposition of the existing land use on the hazard map, combined with knowledge of disaster event damage, can provide a map of current landslide hazard risk. By adjusting susceptibility of present land use in respect of proposed mitigating measures, an appreciation of costs and benefits of the measures can also be achieved.

Figure 6.6 Buildings sited within a debris flow run-out zone, Wairarapa, New Zealand

Landslide hazard maps are also used to control land development and land use. Commonly, zones are assigned to unrestricted use, restricted use, areas of avoidance or areas of qualified use where either more detailed stability investigations are required or where protective measures must be taken.

Table 6.3 **Landslide Probability Classification** (after Crozier 1984)

Class VI	Slopes with active landslides. Material is continually moving, and landslide forms are fresh and well-defined. Movement may be continuous or seasonal.	
Class V	Slopes frequently subject to new or renewed landslide activity. Movement is not a regular, seasonal phenomenon. Triggering of landslides results from events with recurrence intervals of up to five years.	
Class IV	Slopes infrequently subject to new or renewed landslide activity. Triggering of landslides results from events with recurrence intervals greater than five years.	
Class III	Slopes with evidence of previous landslide activity but which have not undergone movement in the preceding 100 years	
	Subclass IIIa	Erosional forms still evident
	Subclass IIIb	Erosional forms no longer present - activity indicated by landslide deposits.
Class II	Slopes which show no evidence of previous landslide activity but which are considered likely to develop landslides in the future. Landslide potential indicated by stress analysis, analogy with other slopes or by analysis of stability factors; several subclasses may be defined.	
Class I	Slopes which show no evidence of previous landslide activity and which by stress analysis, analogy with other slopes or by analysis of stability factors are considered highly unlikely to develop landslides in the foreseeable future.	

APPENDICES

SOME DOCUMENTED LANDSLIDE DISASTERS CAUSING 50 OR MORE FATALITIES

Refer-ence	Location	Year	Number of deaths	Principal active factor
12	Torrent d'Yvorne Switzerland	1584	120	Earthquake precipitation
1	Pleurs Switzerland	1618	2 430	Quarrying
2	Huaraz, Peru	1725	1 500	Earthquake
2	Pueblo de Ancash, Peru	1725	1 500	Earthquake
1	Rossberg Switzerland	1806	457	Foot slope erosion
1	Elm Switzerland	1881	115	Quarrying
1	Verdal Valley Norway	1893	111	Stream erosion and liquefaction
1	Turtle Mountain Canada	1903	76	Mining
1	Kansu, China	1920	10 000	Earthquake
2	Quebrada de Cojup, Peru	1941	5 000	Glacial lake burst

Refer-ence	Location	Year	Number of deaths	Principal active factor
3	Kure, Japan	1945	1 154	Cyclonic rainstorm
1	Tangiwai, New Zealand	1953	154	Volcanic melting of snow
1	Santos, Brazil	1956	100	Rainstorm
4	Southwest of Tokyo, Japan	1958	1 100	Cyclonic rainstorm
5	Rupanco region, China	1960	210	Earthquake
2	Ranrachirca Peru	1962	4 000- 5 000	Ice and rock avalanche
1	Vaiont Gorge Italy	1963	2 117	Variation in groundwater levels
6	Hong Kong Is China	1966	64	Rainstorm
1	Aberfan, Wales	1966	144	Impounded spring water
7	Rio de Janeiro Brazil	1966	1 000	Rainstorm
8	Rio de Janeiro Brazil	1967	1 700	Rainstorm
2	Yungay, Peru	1970	23 000	Earthquake
1	Kwun Tong Hong Kong	1972	100+	Rainstorm
1	Kamijima, Japan	1972	112	Rainstorm
9	Mantaro River Peru	1974	450	High snow year
10	Guatemala City Guatemala	1976	240	Earthquake

Refer- ence	Location	Year	Number of deaths	Principal active factor
1	Pahirikhet Nepal	1976	150	Monsoon rain.
11	Gyansu, Uttar Pradesh, India	1980	150	Rainstorm
11	East of Jakarta Indonesia	1980	100	Rainstorm
11	Mt Semeru	1981	500	Rainstorm
11	Albay Province Philippines	1981	140	Rainstorm
11	Yanacocha, Peru	1981	70	-
11	North Sumatra, Indonesia	1982	50	Rainstorm
11	West of Mon- rovia, Liberia	1982	200	Mine waste collapse
11	Yacitan and Cashipampa, Peru	1983	233 killed 300 missing	Rainstorm
11	Mt Sale Dongxiang, China	1983	277	High, spring water flow
11	Bogota Columbia	1983	150	Rainstorm
11	Sikkim	1983	67	Monsoon rain
11	Western Nepal	1983	186	Rainstorm

References
1   Waltham 1978
2   Zapata Luyo 1977
3   Coates 1977
4   Kawaguchi et al. 1959
5   Wright and Mella 1963
6   So 1971
7   De Meis and Da Silva 1968
8   Jones 1973
9   Lee and Duncan 1975
10 Harp et al. 1981
11 UNDRO 1980-1983
12 Alexander 1983

215

FACTORS INDICATING POTENTIAL STABILITY
CONDITIONS (adapted from Cooke and
Doornkamp 1974 and Crozier 1984)

Factor	Potentially stable	Potentially unstable
Relief		
Valley depth	Small	Very great
Slope steepness	Low	Very steep
Cliffs	Absent	Present
Height difference between valleys	Small	Very great
Valley-side shape	Convex	Concave
Drainage		
Drainage density	Low	Very high
River gradient	Gentle	Very steep
Slope undercutting	None	Very severe
Concentrated seepage	Absent	Present
Standing water	Absent	Present with rapid drainage
Recent incision	Absent	Large
Porewater pressure fluctuation	Low	Very high
Bedrock		
Joint density	Low	Very high
Joint openings	Tight	Wide
Joint fillings	Soft	Hard
Joint wedging (by vegetation and soil)	Absent	Present
Tension cracks	Absent	Present
Strike direction of structural discontinuities with respect to strike of slope	Right-angles	Parallel
Dip angle of structural discontinuities with respect to slope angle	Greater	Lower
Competent over incompetent strata	Absent	Present
Degree of weathering	None	High
Compressive strength	High	Very low
Glauconite in contact with pelitic rocks	Absent	Present
Existing landslides	None	Many
Activity of existing landslides	None	High

Factor	Potentially stable	Potentially unstable
Regolith		
Site	Flat	Steep slope
Coherent over incoherent beds	Absent	Present
Depth	Small	Very large
Shear strength	High	Very low
Plastic index	Low	High
Liquid limit	High	Low
Activity number	Low	High
Desiccation cracks	Absent	Present
Tension cracks	Absent	Present
Permeable over impermeable beds	Absent	Present
Nature of regolith/bedrock contact	Gradational	Abrupt
Subsurface depressions and drainage lines	Absent	Many
Existing landslides	None	Many
Activity of existing landslides	None	High
Earthquake zone		
Tremors felt	None	Many
Felt intensity	Low	High
Proximity to fault	Close	Far
Paleo-features		
Fossil solifluction lobes and sheets	Absent	Many
Fossil gullies	Absent	Many
Previous landslides	Absent	Many
Deep weathering	None	Much
Climate		
Rainstorms	Low intensity	High intensity
Total rainfall variability	Low	High
Drought episodes	Uncommon	Common
Snow-cover melt	Slow	Rapid
Freeze/thaw cycles	Few	Many
Cyclones	Few	Many
Vegetation		
Hydrophytic plants	Absent	Present
Potassium-demanding plants (for example Horsetail)	Absent	Present

Factor	Potentially stable	Potentially unstable
Vegetation (continued)		
Biomass on steep, jointed slopes	Low	High
Biomass on other slopes	High	Low
Recent alteration to biomass	None	Much
Man-made features		
Excavation-depth	Small	Large
Excavation-position	Top of slope	Toe of slope
Fill-compaction	High	None
Reservoir	Absent	Present
Drainage diversion across hillside	Absent	Present
Fluctuation of reservoir level	Small	Large
Loading on upper slope	None	Much
Maintenance of control structures	High	None

REFERENCES

Adams, J E 1978: 'Late Cenozoic erosion in New Zealand.' PhD thesis, Geology Department, Victoria University of Wellington.

Adams, J E 1979: Sediment loads of North Island rivers, New Zealand - a reconnaissance. Journal of Hydrology (New Zealand) 18(1): 36-48.

Ahnert, F 1966: Zur rolle der elektronischen rechenmaschine und des mathematischen modells in der geomorphologie. Geograph. Z. 54: 118-33.

Ahnert, F 1967: The role of the equilibrium concept in the interpretation of landforms of fluvial erosion and deposition. Slopes Commission Report 6: 88-101.

Alexander, D 1983: 'God's handy-worke in wonders' - landslide dynamics and natural hazard implications of a Sixteenth Century disaster. Professional Geographer 35(3): 314-23.

Alfors, J T; Burnett, J L; Gay, T E 1973: Urban geology master plan for California: the nature, magnitude and costs of geologic hazards in California and recommendations for their mitigation. California Division of Mines and Geology Bulletin 198.

Barata, F E 1969: Landslides in the tropical region of Rio de Janeiro. Proceedings of the 7th International Conference on Soil Mechanics and Foundation Engineering, Mexico City, 2: 507-16.

Beaty, C B 1959: Slope retreat by gullying. Geological Society of America Bulletin 70: 1479-82.

Benson, W N 1940: Landslides and allied features in the Dunedin District in relation to geological structure, topography, and engineering. Transactions and Proceedings of the Royal Society of New Zealand 70(3): 249-63.

Bergstrom, F W; Schumm, S A 1981: Episodic behaviour in badlands. Proceedings of Symposium on Erosion and Sediment Transport in Pacific Rim Steeplands, Christchurch, New Zealand 1981. Edited by T R H Davies and A J Pearce. International Association of Hydrological Sciences. Publication 132: 478-92.

Billings, M P 1972: Structural Geology (3rd edition). Prentice-Hall, New Jersey.

Birot, P 1962: Contribution a l'Etude de la Desagregation des Roches. Paris CDU.

Birot, P 1968: The Cycle of Erosion in Different Climates. University of California Press, Berkley.

Bishop, A W 1955: The use of the slip circle in the stability analysis of slopes. Geotechnique 5(1): 7-77.

Bishop, A W 1973: The stability of tips and spoil heaps. Quarterly Journal of Engineering Geology 6: 335-76.

Bishop, A W; Morgenstern, N 1960: Stability coefficients for earth slopes. Geotechnique 4: 49-69.

Bishop, R G 1977: Report on the Storm of 20 December 1976. Wellington Regional Water Board, Wellington, New Zealand.

Blong, R J 1973: A numerical classification of selected landslides of the debris slide-avalanche-flow type. Engineering Geology 7: 99-114.

Brabb, E E 1982: 'Preparation and use of a landslide susceptibility map for a county near San Francisco, California.' In: Landslides and Mudflows: Reports of Alma-Ata International Seminar, October 1981. UNESCO/UNEP. Centre of International Projects, GKNT, Moscow: 407-19.

Brabb, E E 1984: 'Innovative approaches to landslide hazard and risk mapping.' Preprint of the International Symposium on Landslides, Toronto, Canada, September, 1984.

Broch, E; Franklin, J A 1972: The point-load strength test. International Journal of Rock Mechanics and Mining Science 9: 669-97.

Brunsden, D 1973: The application of system theory to the study of mass movement. Geologica Applicata e Idrogeologia 8(1): 185-207.

Brunsden, D; Thornes, J B 1979: Landscape sensitivity and change. Transactions Institute of British Geographers (New Series) 4(4): 463-84.

Bryan, K 1940: Gully gravure - a method of slope retreat. Journal of Geomorphology 3: 89-107.

Budel, J 1944: Die morphologischen wirkugen des eiszeitklimas im gletscherfreien gebiete. Geol. Rundschau 34: 482-519.

Bunting, B T 1964: Slope development and soil formation on some British sandstones. Geographical Journal 130: 506-12.

Burroughs, E R; Thomas, B R 1977: Declining root strength in Douglas fir after felling as a factor in slope stability. United States Department of Agriculture Forest Service Research Paper INT-190.

Caine, N 1980: The rainfall intensity-duration control of shallow landslides and debris flows. Geografiska Annaler 62A(1-2): 23-7.

Campbell, D A 1951: Types of erosion prevalent in New Zealand. Association Internationale d'Hydrologie Scientifique, Assemblee Generale de Bruxelles. Tome II: 82-95.

Campbell, R H 1975: Soil slips, debris flows and rainstorms in the Santa Monica Mountains and vicinity, Southern California. United States Geological Survey Professional Paper 851.

Capper, P L; Cassie, W F 1969: The Mechanics of Engineering Soils. John Wiley and Sons, New York.

Carson, M A 1969: Models of hillslope development under mass failure. Geographical Analysis 1: 76-100.

Carson, M A; Kirkby, M J 1972: Hillslope form and process. Cambridge University Press.

Carson, M A; Petley, D J 1970: The existence of threshold hillslopes in the denudation of the landscape. Transactions of the Institute of British Geographers 49: 71-95.

Casagrande, A 1948: Classification and identification of soils. Transactions of the American Society of Civil Engineers 113: 901-92.

Cato, I 1982: 'The landslide at Tuve 1977 and the complex origin of clays in south-western Sweden.' In: Landslides and Mudflows: Reports of Alma-Ata International Seminar, October 1981. UNESCO/UNEP. Centre of International Projects, GKNT, Moscow: 279-93.

Chinn, T J 1979: How wet is the wettest of the wet West Coast? New Zealand Alpine Journal 32: 35-7.

Clark, C 1983: Flood. Time-Life Books, Amsterdam.

Coates, D F 1964: Classification of rock for rock mechanics. International Journal for Rock Mechanics and Mining Science 1: 421-9.

Coates, D R 1977: 'Landslide perspectives.' In: Landslides: Reviews in Engineering Geology 3. Edited by D R Coates. Geological Society of America, Boulder, Colorado: 3-28.

Conacher, A J 1974: Salt scald. A West Australia case study in rehabilitation. Australian Science and Technology 11(8): 14-6.

Connell, D C; Tombs, J M C 1971: The crystallisation pressure of ice - a simple experiment. Journal of Glaciology 10: 312-5.

Cooke, R U; Doornkamp, J C 1974: Geomorphology in Environmental Management. Clarendon, Oxford.

Cooke, R U; Smalley, I J 1968: Salt weathering in deserts. Nature 220: 1226-7.

Cotton, C A 1958: Alternating Pleistocene morphogenic systems. Geological Magazine 95(2): 123-37.

Cotton, C A; Te Punga, M T 1955: Solifluxion and periglacially modified landforms of Wellington, New Zealand. Transactions of the Royal Society of New Zealand 82(5): 1001-31.

Coulomb, C A 1776: Essai sur une application des regles des maximis et minimis a quelques problemes de statique relatifs a l'architecture. Memorandum Academie Royale, Pres. Div. Sav. 7: 343-82.

Crozier, M J 1972: 'Some problems in the correlation of landslide movement and climate.' In: International Geography 1972/La Geographie Internationale, Part 1, Paper No PO148: 90-3.

Crozier, M J 1973: Techniques for the morphometric analysis of landslips. Zeitschrift fur Geomorphologie 17(1): 78-101.

Crozier, M J 1982a: 'A technique for predicting the probability of mudflow and rapid landslide occurrence.' In: Landslides and Mudflows: Reports of Alma-Ata International Seminar, October 1981. UNESCO/UNEP. Centre of International Projects, GKNT, Moscow: 420-30.

Crozier, M J 1982b: The character of natural hazards in different physical and social settings. Proceedings of the Eleventh New Zealand Geography Conference, Wellington 1981: 106-8.

Crozier, M J 1983: The mass movement regime: recent history of landslide activity in the Wairarapa hill country, New Zealand. New Zealand Geomechanics News 27: 3-9.

Crozier, M J 1984: 'Field assessment of slope instability.' In: Slope Instability. Edited by D Brunsden and D Prior. John Wiley and Sons: 103-42.

Crozier, M J; Eyles, R J 1980: Assessing the probability of rapid mass movement. Third Australia-New Zealand Conference on Geomechanics, Wellington, 1980 - Volume 2. New Zealand Institution of Engineers, Proceedings of Technical Groups 6(1G): 2.47-2.53.

Crozier, M J; Eyles, R J; Marx, S L; McConchie, J A; Owen, R C 1980: Distribution of landslips in the Wairarapa hill country. New Zealand Journal of Geology and Geophysics 23: 575-86.

Crozier, M J; Howorth, R; Grant I J 1981: Landslide activity during cyclone Wally, Fiji: a case study of the Wainitubatolu Catchment. Pacific Viewpoint 22(1): 69-80.

Crozier, M J; McConchie, J A; Owen, R C; Eyles, R J 1982: Mass Movement Erosion - Wairarapa. Department of Geography, Victoria University of Wellington.

Culmann, C 1866: Graphische Statik. Zurich.

Cumberland, K B 1947: Soil Erosion in New Zealand. Whitcombe and Tombs Ltd.

Cunningham, A; Arnott, W B 1964: Observations following a heavy rainfall on the Rimutaka Range. Journal of Hydrology (New Zealand) 3(2): 15-24.

Cunningham, M J 1971: A mathematical model of the physical processes of an earthflow. Proceedings of the Seminar on Catchment Control. New Zealand Association of Soil Conservators: 310-8.

Da Costa Nunes, A J 1969: Landslides in soils of decomposed rock due to intense rainstorms. Proceedings of the Seventh International Conference on Soil Mechanics and Foundation Engineering 2: 547-54.

Dahl, R 1966: Block fields, weathering pits and Tor-like forms in the Narvik mountains, Nordland, Norway. Geografisker Annaler 48: 55-85.

Dansereau, P 1957: Biogeography: an Ecological Perspective. Ronald Press, New York.

Davis, W M 1896: Plains of marine and sub-aerial denudation. Bulletin Geological Society of America 7: 377-98.

Davis, W M 1902: Base-level, grade, and peneplain. Journal of Geology 10: 77-111.

Deere, D U; Miller, R P 1966: Engineering classification and index properties for intact rock. Technical Report No AFNL-TR-65-116 Air Force Weapons Laboratory, New Mexico.

De Meis, M R M; Da Silva, J X 1968: Movements de mass recents a Rio de Janeiro: une etude de geomorphologie dynamique. Revue de Geomorphologie Dynamique 18(4): 145-51.

Dibble, R R; Neall, V E 1984: 'Volcanic hazards in New Zealand.' In: Natural Hazards in New Zealand. Edited by I Speden and M J Crozier. New Zealand National Commission for UNESCO, Wellington: 332-74.

Dietrich, W E; Dunne, T 1978: Sediment budget for a small catchment in mountainous terrain. Zeitschrift fur Geomorphologie Supplement 29: 191-206.

Dietrich, W E; Dunne, T; Humphrey, N S; Reid, L M 1982: 'Construction of sediment budgets for drainage basins.' In: Sediment Budgets and Routing in Forested Drainage Basins. Edited by F J Swanson, R J Janda, T Dunne, D N Swanston. United States Department of Agriculture, Forest Service, General Technical Report PNW-141: 5-23.

Durgin, P B 1977: 'Landslides and weathering of granitic rocks.' In: Landslides: Reviews in Engineering Geology 3. Edited by D R Coates. Geological Society of America, Boulder, Colorado: 127-34.

East, T J 1978: Mass movement landforms in Baroon Pocket, south-east Queensland: a study of form and process. Queensland Geographical Journal (3rd Series) 4: 37-67.

Eyles, G O 1983: The distribution and severity of present soil erosion in New Zealand. New Zealand Geographer 39(1): 12-28.

Eyles, R J 1971: Mass movement in Tangoio Conservation Reserve, Hawkes Bay. Earth Science Journal 5(2): 79-91.

Eyles, R J 1979: Slip-triggering rainfalls in Wellington City, New Zealand. New Zealand Journal of Science 22: 117-21.

Eyles, R J; Crozier, M J; Wheeler, R H 1978: Landslips in Wellington City. New Zealand Geographer 34(2): 58-74.

Eyles, R J; Eyles, G O 1982: Recognition of storm damage events. Proceedings of Eleventh New Zealand Geography Conference, Wellington 1981: 118-23.

Fahey, B D 1973: An analysis of diurnal freeze-thaw and frost heave cycles in the Indian Peaks Region of the Colorado Front Range. Arctic and Alpine Research 5(3): 269-81.

Fahey, B D 1983: Frost action and hydration as rock weathering mechanisms on schist: a laboratory study. Earth Surface Processes and Landforms 8(6): 535-45.

Fairbridge, R W 1978: 'Global cycles and climate.' In: Climatic Change and Variability. Edited by A B Pittock, L A Frakes, D Jenssen, J A Peterson, J W Zillman. Cambridge University Press: 200-11.

Fellenius, W 1927: Erdstatische Berechnungen (revised 1939). W Ernst and Sohn, Berlin.

Fiksdal, A J 1974: A Landslide Survey of Stequaleho Creek Watershed. Supplement to Final Report FRI-UW-7404, Fisheries Research Institute, University of Washington.

Flaccus, E 1959: Revegetation of landslides in the White Mountains of New Hampshire. Ecology 40: 692-703.

Fleming, R W; Taylor, F A 1980: Estimating the cost of landslide damage in the United States. Geological Survey Circular 832.

Fookes, P G; Horswill, P 1970: Discussion on engineering grade zones. Proceedings of the Conference on In-situ Investigations in Soils and Rock, British Geotechnical Society, London: 53-7.

Fujihara, K 1978#: A study of landslides by aerial photographs. Hokkaido University Agriculture Department Experiment Plantation Research Report 27(2). (#Date of translation by R R Ziemer, J M Arata, USDA Forest Service)

Goudie, A S 1974: Further experimental investigation of rock weathering by salt and other mechanical weathering processes. Zeitschrift fur Geomorphologie Supplement 21: 1-12.

Goudie, A S; Watson, A 1984: Rock block monitoring of rapid salt weathering in southern Tunisia. Earth Surface Processes and Landforms 9(1): 95-8.

Grant, G E; Crozier, M J; Swanson, F J (in press): An approach to evaluating off-site effects of timber harvest activities on channel morphology. Proceedings of Symposium on the Effects of Forest Land Use on Erosion and Slope Stability, May 1984, Honolulu Hawaii.

Grant, P J 1983: Recently increased erosion and sediment transport rates in the Upper Waipawa River Basin, Ruahine Range, New Zealand. Soil Conservation Centre, Aokautere, Publication 5, Water and Soil Division, Ministry of Works and Development, New Zealand.

Grant-Taylor, T L 1964: Stable angles in Wellington greywacke. New Zealand Engineering 19(4): 129-30.

Gregory, K J; Brown, E H 1966: Data processing and the study of land form. Zeitschrift fur Geomorphologie 10: 237-63.

Hallin, W E 1967: Soil moisture and temperature trends in cutover and adjacent old-growth Douglas fir timber. United States Department of Agriculture Forest Service Research Note PNW-56.

Hansbo, S 1957: A new approach to the determination of the shear strength of clay by the fall-cone test. Swedish Geotechnical Institute Proceedings No 14.

Hansen, A 1984: 'Landslide hazard analysis.' In: Slope Instability. Edited by D Brunsden and D B Prior. John Wiley and Sons: 523-602.

Hansen, W R 1966: Effects of the earthquake of March 27, 1964, at Anchorage, Alaska. United States Geological Survey Professional Paper 542-A.

Harp, E L; Wilson, R C; Wieczorek, G F 1981: Landslides from the February 4, 1976, Guatemala earthquake. United States Geological Survey Professional Paper 1204-A.

Hartsog, W S; Martin, G H 1974: Failure conditions in infinite slopes and resulting soil pressures. United States Department of Agriculture, Forest Service Research Paper INT-149.

Hawley, J G 1980: 'Introduction.' In: Workshop on the Influence of Soil Slip Erosion on Hill Country Pastoral Productivity. Aokautere Science Centre, Internal Report 21, Water and Soil Division, Ministry of Works and Development, New Zealand: 4-6.

Hawley, J G 1981: The literates and the numerates. New Zealand Geomechanics News 23: 3-6.

Hawley, J G 1984: 'Slope instability in New Zealand.' In: Natural Hazards in New Zealand. Edited by I Speden and M J Crozier. New Zealand National Commission for UNESCO, Wellington: 88-133.

Hellberg, M 1984: 'Insurance and natural hazards: the response of New Zealand governments.' In: Natural Hazards in New Zealand. Edited by I Speden and M J Crozier. New Zealand National Commission for UNESCO: 446-65.

Henkel, D J; Skempton, A W 1954: A landslide at Jackfield, Shropshire in an over-consolidated clay. Proceedings of the Conference on Stability of Earth Slopes, Stockholm 1: 90-101.

Hodder, A P W 1976: Cavitation-induced nucleation of ice: a possible mechanism for frost-cracking in rocks. New Zealand Journal of Geology and Geophysics 19: 821-6.

Hoek, E 1970: Estimating the stability of excavated slopes in open-cast mines. Transactions, Section A, Institute of Mining and Metallurgy 79: 109-32.

Hoek, E; Bray, J 1977: Rock Slope Engineering (2nd Edition). Institute of Mining and Metallurgy, London.

Hoek, E; Londe, P 1974: Surface workings in rock. Proceedings 3rd Congress, International Society for Rock Mechanics, Denver IA: 613-52.

Holtz, W G; Gibbs, H J 1956: Engineering properties of expansive clay. Transactions of the American Society of Civil Engineers 121: 641-77.

Hradek, M 1977: Distribution of fine-grained products of Pleistocene frost weathering in the Ceska vysocina and their geomorphological importance. Z. Geografickeho Ustavu, CSAV 14: 92-6.

Hutchinson, J N 1965: 'The stability of cliffs composed of soft rocks, with particular reference to the coasts of south-east England.' PhD Dissertation, Cambridge University.

Hutchinson, J N 1968: 'Mass movement.' In: The Encyclopedia of Geomorphology. Edited by R W Fairbridge. Reinhold: 688-96.

Hutchinson, J N 1975: The response of London Clay cliffs to differing rates of toe erosion. Geologie e Idrogeologia 7(1) Bari: 222-39.

Hutchinson, J N 1976: Coastal landslides in cliffs of Pleistocene deposits between Cromer and Overstrand, Norfolk, England. Lauritis Bjerrum Memorial Volume Norwegian Geotechnical Institute, Oslo: 155-82.

Hutchinson, J N 1977: 'General, largely morphological classification of mass movements on slopes - November 1977.' Unpublished Teaching Hand-out, Imperial College, London.

Hutchinson, J N 1978: 'A geotechnical classification of landslides (After A W Skempton and J N Hutchinson).' Unpublished Teaching Hand-out, Imperial College, London.

Hutchinson, J N 1979: Various forms of cliff instability arising from coast erosion in south-east England. Fjellsprengningsteknikk Bergmekanikk – Geoteknikk. Norwegian Geotechnical Society: 19.1-19.32.

Hutchinson, J N 1982: 'Damage to slopes produced by seepage erosion in sands.' In: Landslides and Mudflows: Reports of Alma-Ata International Seminar, October 1981. UNESCO/UNEP. Centre of International Projects, GKNT, Moscow: 250-68.

Hutchinson, J N; Bhandari, R K 1971: Undrained loading, a fundamental mechanism of mudflows and other mass movements. Geotechnique 21(4): 353-8.

Hutchinson, J N; Gostelow, T P 1976: The development of an abandoned cliff in London Clay at Hadleigh, Essex. Philosophical Transactions Royal Society London A 283: 557-604.

Iida, T; Okunishi, K 1983: Development of hillslopes due to landslides. Zeitschrift fur Geomorphologie 46: 67-77.

Ikeya, H 1976: Introduction to Sabo Works: the Preservation of Land Against Sediment Disaster. The Japan Sabo Association, Tokyo.

James, I L 1973: Mass movement in the upper Pohangina Catchment, Ruahine Range. Journal of Hydrology (New Zealand) 12(2): 92-102.

Janbu, N 1973: 'Slope stability computations.' In: Embankment and Dam Engineering. Edited by R C Hirshfeld and S J Poulos. Wiley: 47-86.

Johnson, A M 1970: Physical Processes in Geology. Freeman Cooper.

Jones, F O 1973: Landslides of Rio de Janeiro and the Serra das Araras escarpment, Brazil. United States Geological Survey Professional Paper 697.

Jones, J A A 1978: Soil pipe networks: distribution and discharge. Cambria 5(1): 1-21.

Kassiff, G; Livneh, M; Wiseman, G 1969: Pavements on Expansive Clays. Jerusalem Academic Press.

Kawaguchi, T; Namba, S; Takiguchi, K; Kono, R; Kishioka, T 1959: Landslides and soil losses at the mountain districts of Izu Peninsula in the flood of 1958 and their control. Japan Forest Experiment Station Bulletin 117: 83-120.

Keefer, D K 1984: Landslides caused by earthquakes. Geological Society of America Bulletin 95(4): 406-21.

Kelly, W C; Zumberge, J H 1961: Weathering of a quartz diorite at Marble Point, McMurdo Sound, Antarctica. Journal of Geology 69(4): 433-46.

Kelsey, H M 1980: A sediment budget and an analysis of geomorphic process in the Van Duzen River basin, north coastal California. Geological Society of America Bulletin, Part II 91: 1119-216.

Kelsey, H M; Madej, M A; Pitlick, J; Coghlan, M; Best, D; Belding, R; Stroud, P 1981: Sediment sources and sediment transport in Redwood Creek basin: a progress report. Redwood National Park Research and Development, Technical Report 3.

Kerrison, G C 1981: 'Tunnel gully erosion in an east Wairarapa hill country catchment.' BSc (Hons) Dissertation, Department of Geography, Victoria University of Wellington.

Kirkby, M J 1971: 'Hillslope process-response models based on the continuity equation.' In: Slopes Form and Process. Edited by D Brunsden. Institute of British Geographers Special Publication 3: 15-30.

Kirkpatrick, W M 1965: Effects of grain size and grading on shearing behaviour of granular materials. Proceedings of 6th International Conference on Soil Mechanics and Foundation Engineering 1: 273-7. University of Toronto Press.

Kohler, M A; Linsley, R K 1951: Predicting the runoff from storm rainfall. United States Weather Bureau Research Paper 34.

Langenheim, J H 1956: Plant succession on a subalpine earthflow in Colorado. Ecology 37: 301-17.

Lawrence, J H 1981: 'Urban capability as part of land use planning.' In: Geomechanics in Urban Planning. Institution of Professional Engineers, New Zealand 9(2G): 328-36.

Lee, I K; White, W; Ingles, O G 1983: Geotechnical Engineering. Pitman.

Lee, K L; Duncan, J M 1975: Landslide of April 25, 1974 on the Mantaro River, Peru. National Academy of Sciences, Washington DC.

Lehre, A K 1981: 'Sediment mobilization and production from a small Coast Range catchment: Lone Tree Creek, Marin Co., California.' PhD Thesis, University of California, Berkeley.

Lehre, A K 1982: 'Sediment budget of a small Coast Range drainage basin in North-Central California.' In: Sediment Budgets and Routing in Forested Drainage Basins. Edited by F J Swanson, R J Janda, T Dunne, D N Swanston. United States Department of Agriculture, Forest Service, General Technical Report PNW-141: 67-77.

Lensen, G J; Vella, P 1971: The Waiohine River faulted terrace sequence. Royal Society of New Zealand Bulletin 9: 117-9.

Lo, K Y 1965: Stability of slopes in anisotropic soils. Journal of Soil Mechanics and Foundations Division ASCE 91(SM4): 85-106.

Lumb, P 1962: The properties of decomposed granite. Geotechnique 12: 226-43.

Lumb, P 1965: The residual soils of Hong Kong. Geotechnique 15: 180-94.

Lumb, P 1966: The variability of natural soils. Canadian Geotechnical Journal 3: 74-97.

Lumb, P 1974: 'Application of statistics in soil mechanics.' In: Soil Mechanics - New Horizons. Edited by I K Lee. Newnes-Butterworths: 44-111.

Lutz, H J 1960: Movement of rocks by uprooting of forest trees. American Journal of Science 258: 752-6.

Lyell, C 1853: A Manual of Elementary Geology. Appleton and Co, New York.

Lynn, I H; Eyles, G O 1981: 'The extent of tunnel gully erosion in New Zealand.' Paper presented at New Zealand Institute of Agricultural Science Convention, 1981.

McConchie, J A 1977: 'The geomorphic and hydrological response of the Stokes Valley Catchment to the 20th December 1976 storm.' BSc (Hons) Dissertation, Department of Geography, Victoria University of Wellington.

McGreevy, J P 1982: 'Frost and salt' weathering: further experimental results. Earth Surface Processes and Landforms 7(5): 475-88.

McSaveney, M J 1978: 'The magnitude of erosion across the Southern Alps.' Paper presented at Erosion Assessment and Control Conference, Christchurch, New Zealand.

Madej, M A 1982: 'Sediment transport and channel changes in an aggrading stream in Puget Lowland, Washington.' In: Sediment Budgets and Routing in Forested Drainage Basins. Edited by F J Swanson, R J Janda, T Dunne, D N Swanston. United States Department of Agriculture, Forest Service, General Technical Report PNW-141: 97-108.

Mark, A F; Scott, G A M; Sanderson, F R; James, P W 1964: Forest succession on landslides above Lake Thomson, Fiordland. New Zealand Journal of Botany 2: 60-89.

Marron, D C (in press); Colluvium in bedrock hollows on steep slopes, Redwood Creek drainage basin, Northwestern California. Catena.

Martin, G R; Millar, P J 1974: Stability of slopes in weathered and jointed rocks. Proceedings of the Symposium on Stability of Slopes in Natural Ground, Nelson, November 1974. New Zealand Geomechanics Society: 7.1-7.14.

Means, R E; Parcher, J V 1963: Physical Properties of Soils. Charles E Merrill Books, Columbus.

Meehan, W R 1974: Fish habitat and timber harvest in southeast Alaska. Naturalist 25: 28-31.

Ministry of Works and Development 1977: Filed data list, Volume 13 (unpublished). Ministry of Works and Development, New Zealand.

Mohr, O 1914: Abhandlungen aus dem Gebiete der Technischen Mechanik. Sohn, Berlin.

Moon, B P; Selby, M J 1983: Rock mass strength and scarp forms in southern Africa. Geografiska Annaler 65A(1-2): 135-45.

Morgenstern, N R; Price, V E 1965: The analysis of the stability of general slip surfaces. Geotechnique 15(1): 79-93.

Morrison, P H 1975: 'Ecological and geomorphological consequences of mass movements in the Alder Creek watershed and implications for forest land management.' BS Honours Thesis, University of Oregon, Eugene.

Mosley, M P 1978: 'Erosion in the protection forests.' Paper presented at Erosion Assessment and Control Conference, Christchurch, New Zealand.

Moss, M R; Rosenfeld, C L 1978: Morphology, mass wasting and forest ecology of a Post-Glacial re-entrant valley, Niagara Escarpment. Geografiska Annaler 60A(3-4): 161-74.

New Zealand Geomechanics Society 1980: Draft Method of Soil Description for Engineering Use. New Zealand Geomechanics Society.

New Zealand Geomechanics Society 1983: New Zealand Geomechanics Society submission to the Ministry of Works and Development Committee to Inquire into the Wheao Canal failure. New Zealand Geomechanics News 26 (June): 20-4.

New Zealand Law Society 1984: 'Law and natural hazards.' In: Natural Hazards in New Zealand. Edited by I Speden and M J Crozier. New Zealand National Commission for UNESCO, Wellington: 433-45.

Nilsen, T H; Taylor, F A; Dean, R M 1976: Natural conditions that control landsliding in the San Francisco Bay region - an analysis based on data from the 1968-69 and 1972-73 rainy seasons. United States Geological Survey Bulletin 1424.

Northey, R D; Hawley, J G; Barker, P R 1974: Classification and mechanisms of slope failure in natural ground. Proceedings of the Symposium on Stability of Slopes in Natural Ground, Nelson, November 1974. New Zealand Geomechanics Society: 3.1-3.8

O'Byrne, T N 1967: A correlation of rock types with soils, topography and erosion in the Gisborne-East Cape region. New Zealand Journal of Geology and Geophysics 10(1): 217-31.

Okimura, T 1983: Rapid mass movement and groundwater level movement. Zeitschrift fur Geomorphologie Supplement 46: 35-54.

O'Loughlin, C L 1972: 'An investigation of the stability of steepland forest soils in the Coast Mountains, southwest British Columbia.' PhD Thesis, University of British Columbia, Vancouver.

O'Loughlin, C L 1974: The effects of timber removal on the stability of forest soils. Journal of Hydrology (New Zealand) 13(2): 121-34.

O'Loughlin, C L; Gage, M 1975: 'A report on the status of slope erosion on selected steep areas, West Coast beech project area.' Unpublished New Zealand Forest Service Report, Christchurch.

O'Loughlin, C L; Pearce, A J 1976: Influence of Cenozoic geology on mass movement and sediment yield response to forest removal, north Westland, New Zealand. Bulletin of the International Association of Engineering Geology 14: 41-6.

O'Loughlin, C L; Pearce, A J 1982: 'Erosion processes in the mountains.' In: Landforms of New Zealand. Edited by J M Soons and M J Selby. Longman Paul, Auckland: 67-79.

O'Loughlin, C L; Ziemer, R R 1982: 'The importance of root strength and deterioration rates upon edaphic stability in steepland forests.' In: Carbon Uptake and Allocation in Subalpine Ecosystems as a Key to Management. Proceedings of International Union of Forest Research Organisations, August 1982, Oregon State University, Corvallis, Oregon: 70-8.

Owen, R C 1981: Soil strength and microclimate in the distribution of shallow landslides. Journal of Hydrology (New Zealand) 20(1): 17-26.

Pain, C F; Bowler, J M 1973: Denudation following the November 1970 earthquake at Madang, Papua New Guinea. Zeitschrift fur Geomorphologie 18: 92-104.

Patton, F D 1966: Multiple modes of shear failure in rock. Proceedings of the 1st Congress International Society for Rock Mechanics, Lisbon 1: 509-73.

Penck, W 1953: Morphological Analysis of Landforms. Macmillan, London.

Pender, M J 1971: Some properties of weathered greywacke. Proceedings of the 1st Australia-New Zealand Conference on Geomechanics 1: 423-9.

Penman, H L 1948: Natural evaporation from open water, bare soil and grass. Proceedings of the Royal Society A193: 120-45.

Penner, E 1963: Sensitivity in Leda clay. Nature 197: 347-8.

Petak, W J; Atkisson, A A 1982: Natural Hazard Risk Assessment and Public Policy. Springer-Verlag, New York.

Petley, D J 1984: 'Ground investigation, sampling and testing for studies of slope instability.' In: Slope Instability. Edited by D Brunsden and D B Prior. John Wiley and Sons: 67-101.

Phillips, R W 1971: 'Effects of sediment on the gravel environment and fish production.' In: Forest Land Use and Stream Environment. Oregon State University, Corvallis: 64-74.

Pierson, T C 1977: 'Factors controlling debris-flow initiation on forested hillslopes in the Oregon Coast Range.' PhD Dissertation, University of Washington, Seattle.

Pierson, T C 1982: Classification and hydrological characteristics of scree slope deposits in the northern Craigieburn Range, New Zealand. Journal of Hydrology (New Zealand) 21(1): 34-60.

Pierson, T C 1983: Soil pipes and slope stability. Quarterly Journal of Engineering Geology London 16: 1-11.

Piteau, D R 1971: Geological factors significant to the stability of slopes cut in rock. Proceedings of the Open Pit Mining Symposium. South African Institute of Mining and Metallurgy: 33-5.

Pitty, A F 1971: Introduction to Geomorphology. Methuen & Co Ltd, London.

Price, L W 1972: The periglacial environment, permafrost and man. Commission on College Geography Resource Paper No 14. Association of American Geographers.

Radbruch, D H; Crowther, K C 1973: Map showing areas of estimated relative amounts of landslides in California. United States Geological Survey Miscellaneous Investigations Map 1-747.

Rankine, W J M 1857: On the stability of loose earth. Philosophical Transactions of the Royal Society, London 147: 9-28.

Ravina, I; Zaslavsky, D 1974: The electric double layer as a possible factor in desert weathering. Zeitschrift fur Geomorphologie NF Supplement 21: 13-8.

Rhodes, D C 1968: Landsliding in mountainous humid tropics: a statistical analysis of landmass denudation in New Guinea. Office of Naval Research, Geography Branch, Technical Report 4.

Rice, R M 1982: 'Sedimentation in the chaparral: how do you handle unusual events?' In: Sediment Budgets and Routing in Forested Drainage Basins. Edited by F J Swanson, R J Janda, T Dunne, D N Swanston. United States Department of Agriculture, Forest Service, General Technical Report PNW-141: 39-49.

Rice, R M; Corbett, E S; Bailey, R G 1969: Soil slips related to vegetation, topography, and soils in southern California. Water Resources Research 5(3): 647-59.

Rice, R M; Foggin, G T 1971: Effect of high intensity storms on soil slippage on mountainous watersheds in southern California. Water Resources Research 7(6): 1485-96.

Riddolls, B W; Grocott, G G 1983: A system for improved description of engineering properties of geological material. New Zealand Geomechanics News 27: 42-3.

Riddolls, B W; Perrin, N D 1975: Wellington urban motorway - The Terrace tunnel: engineering geological investigations. New Zealand Engineering 30(8): 221-4.

Robertson, A M 1971: The interpretation of geological factors for use in slope theory. Proceedings of the Open Pit Mining Symposium. South African Institute of Mining and Metallurgy: 55-71.

Ruxton, P B 1958: Weathering and subsurface erosion in granite at the piedmont angle, Balos, Sudan. Geological Magazine 95: 353-77.

Schumm, S A 1963: The disparity between present rates of denudation and orogeny. United States Geological Survey, Professional Paper 454-H.

Schuster, R L 1978: 'Introduction.' In: Landslides Analysis and Control. Edited by R L Schuster and R J Krizek. Transportation Research Board Special Report 176, National Academy of Sciences, Washington DC: 1-10.

Schuster, R L 1982: 'Environmental effects of landslides associated with the 18 May 1980 eruption of Mount St Helens, State of Washington.' In: Landslides and Mudflows: Reports of Alma-Ata International Seminar, October 1981. UNESCO/UNEP. Centre of International Projects, GKNT, Moscow: 69-86.

Seed, H B 1966: A method for earthquake resistant design of earth dams. Journal of Soil Mechanics and Foundations Division, ASCE 92(SM1): 13-41.

Seed, H B; Woodward, R J; Lundgren, R 1964: Fundamental aspects of Atterberg limits. Journal of the Soil Mechanics and Foundations Division, ASCE 90 (SM6)4140: 75-105.

Selby, M J 1976: Slope erosion due to extreme rainfall: a case study from New Zealand. Geografiska Annaler 3(A): 131-8.

Selby, M J 1979: 'Slope stability studies in New Zealand.' In: Physical Hydrology: New Zealand Experience. Edited by D L Murray and P Ackroyd. New Zealand Hydrological Society: 120-34.

Selby, M J 1980: A rock mass strength classification for geomorphic purposes: with tests from Antarctica and New Zealand. Zeitschrift fur Geomorphologie 24(1): 31-51.

Selby, M J 1982a: Controls on the stability and inclinations of hillslopes formed on hard rock. Earth Surface Processes and Landforms 7: 449-67.

Selby, M J 1982b: Rock mass strength and the form of some Inselbergs in the Central Namib Desert. Earth Surface Processes and Landforms 7: 489-97.

Selby, M J 1982c: Hillslope Materials and Process. Oxford.

Sharpe, C F S 1938: Landslides and Related Phenomena. Pageant, New Jersey.

Sheppard, D S; Adams, C J; Bird, G W 1975: Age of metamorphism and uplift in the Alpine Schist Belt, New Zealand. Geological Society of America Bulletin 86: 1147-53.

Sidle, R C; Pearce, A J; O'Loughlin, C L 1985: Soil mass movement - influence of natural factors and land use. American Geophysical Union Water Resources Monograph 11. American Geophysical Union, Washington DC.

Sidle, R C; Swanston, D N 1982: Analysis of a small debris slide in coastal Alaska. Canadian Geotechnical Journal 19: 167-74.

Skempton, A W 1948: The rate of softening in stiff, fissured clays. Proceedings of the 2nd International Conference on Soil Mechanics and Foundation Engineering 2: 50-3.

Skempton, A W 1953: Soil mechanics in relation to geology. Proceedings of Yorkshire Geological Society 29 Pt 1(3): 33-62.

Skempton, A W 1964: The long-term stability of clay slopes. Geotechnique 14: 75-102.

Skempton, A W 1970: First-time slides in overconsolidated clays. Geotechnique 20: 320-4.

Skempton, A W; DeLory, F A 1957: Stability of natural slopes in London Clay. Proceedings of the 4th International Conference on Soil Mechanics and Foundation Engineering 2: 378-81.

Skempton, A W; Hutchinson, J N 1969: Stability of natural slopes and embankment foundations. State of the Art Volume, 7th International Conference on Soil Mechanics and Foundation Engineering, Mexico: 291-340.

Skempton, A W; Northey, R D 1952: The sensitivity of clays. Geotechnique 3: 30-53.

Smalley, I J; Bentley, S P 1980: Towards a general theory of sensitivity. New Zealand Geomechanics News 20: 22-5.

Smith, R D; Hicks, B G 1982: 'Ashland Creek drainage basin sediment budgets.' In: Sediment Budgets and Routing in Forested Drainage Basins. Edited by F J Swanson, R J Janda, T Dunne, D N Swanston. United States Department of Agriculture, Forest Service, General Technical Report PNW-141: 112-3.

Snow, D T 1964: 'Landslides of Cerro Condor-Seneca, Department of Ayacucho, Peru.' In: Engineering Geology Case Histories No 1-5. Edited by P D Trask and G A Kiersch. Geological Society of America, Boulder, Colorado: 243-8.

So, C L 1971: Mass movements associated with the rainstorm of June 1966 in Hong Kong. Institute of British Geographers Transactions 53: 55-65.

Soderblom, R 1969: Salt in Swedish clays and its importance for quick clay formation. Swedish Geotechnical Institute Proceedings No 22.

Soderblom, R 1974: New lines in quick clay research. Swedish Geotechnical Institute Reprints and Preliminary Reports No 55.

Soil and Water 1982: Ruahihi's raging river of rubble: the official report. Soil and Water 18(1): 4-10.

Soil and Water 1983: The whys and wherefores of Wheao. Soil and Water 19(3): 5-22.

Spencer, E 1967: A method of analysis of the stability of embankments assuming parallel inter-slice forces. Geotechnique 17(1): 11-26.

Spencer, E 1968: Effect of tension on stability of embankments. Journal of Soil Mechanics and Foundations Division, ASCE 94(SM5): 1161-73.

Stacey, F D 1969: Physics of the Earth. John Wiley and Sons Inc., New York.

Stevens, G R 1974: Rugged Landscape. A H and A W Reed.

Stewart, J W 1964: Infiltration and permeability of weathered crystalline rocks. United States Geological Survey Bulletin 1133.

Stone, E L; Swank, W T; Hornbeck, J W 1978: 'Impact of timber harvest and regeneration systems on streamflow and soils in the eastern deciduous region.' In: Forest Soils and Land Use. Edited by G T Youngberg. Proceedings of the 5th North American Forest Soils Conference, Fort Collins, Colorado.

Susman, P; O'Keefe, P ; Wisner, B 1983: 'Global disasters, a radical interpretation.' In: Interpretations of Calamity. Edited by K Hewitt. Allen and Unwin: 263-83.

Swanson, F J; Dyrness, C T 1975: Impact of clear-cutting and road construction on soil erosion by landslides in the Western Cascade Range, Oregon. Geology 3(7): 393-6.

Swanson, F J; James, M E 1975: Geology and geomorphology of the H J Andrews Experimental Forest, western Cascades, Oregon. United States Department of Agriculture, Forest Service Research Paper PNW-188.

Swanston, D N 1976: Erosion processes and control methods in North America. The 16th International Union of Forest Research Organisations World Congress Proceedings 1: 251-75.

Swanston, D N 1977: Prediction, prevention and control of landslides on mountainous forest lands. Proceedings of Seminar on Watershed Management, Peshawar, Pakistan, September 1977: 226-49.

Taber, S 1929: Frost heaving. Journal of Geology 37: 428-61.

Taylor, A J W 1983: Hidden victims and the human side of disasters. United Nations Disaster Relief Organisation News, March/April: 6-12.

Taylor, D K; Hawley, J G; Riddolls, B W 1977: Slope Stability in Urban Development. New Zealand Geomechanics Society and Department of Scientific and Industrial Research, Wellington.

Taylor, D W 1948: Fundamentals of Soil Mechanics. John Wiley and Sons, New York.

Taylor, R K; Spears, D A 1970: The breakdown of British coal measure rocks. International Journal of Rock Mechanics and Mining Science 7: 481-501.

Taype, V 1979: Los desastres naturales como problema de la defensa civil. Bol. Soc. Geolog. del Peru 61: 101-11.

Terzaghi, K 1936: The shearing resistance of saturated soils and the angles between the planes of shear. Proceedings of 1st International Conference on Soil Mechanics and Foundation Engineering 1: 54-6.

Terzaghi, K 1943: Theoretical Soil Mechanics. New York.

Terzaghi, K 1944: Ends and means in soil mechanics. Engineering Journal (Canada) 27: 608-15.

Terzaghi, K 1950: Mechanism of landslides. Geological Society of America Berkey Volume: 83-123.

Terzaghi, K 1962: Stability of steep slopes on hard unweathered rock. Geotechnique 12(4): 251-70.

Terzaghi, K; Peck, R B 1967: Soil Mechanics in Engineering Practice. (2nd Edition) John Wiley and Sons, New York.

Thompson, S M; Adams, J 1979: 'Suspended sediment load in some major rivers of New Zealand.' In: Physical Hydrology: New Zealand Experience. Edited by D L Murray and P Ackroyd. New Zealand Hydrological Society: 213-29.

Tomlinson, A I 1977: The Wellington and Hutt Valley flood of 20 December, 1976. New Zealand Meteorological Service Technical Information Circular 154.

Tricart, J; Cailleux, A 1972: Introduction to Climatic Geomorphology. Longman.

Trustrum, N A 1981: Use of sequential aerial photography to assess the long term effect of soil slip erosion on hill country pasture production. New Zealand Association of Soil Conservators, Broadsheet, December: 42-7.

Tsukamoto, Y; Ohta, T; Noguchi, H 1982: 'Hydrological and geomorphological studies of debris slides on forested hillslopes in Japan.' In: Recent Developments in the Explanation and Prediction of Erosion and Sediment Yield. International Association of Hydrological Sciences Publication 137: 89-98.

United Nations 1979: Disaster Prevention and Mitigation, Volume 7: Economic Aspects. Office of United Nations Disaster Relief Co-ordinator, Geneva.

United Nations 1980: Disaster Prevention and Mitigation, Volume 9: Legal Aspects. Office of United Nations Disaster Relief Co-ordinator, Geneva.

UNDRO 1980-83: United Nations Disaster Relief Organisation Newsletters 1980-83. Geneva.

USGS 1982: Goals and tasks of the landslide part of a ground failure hazards reduction programme. United States Geological Survey Circular 880.

Vandre, B C; Swanston, D N 1977: A stability evaluation of debris avalanches caused by blasting. Association of Engineering Geologists, Bulletin 15(4): 205-25.

Varnes, D J 1958: 'Landslide types and processes.' In: Landslides and Engineering Practice. Edited by E B Eckel. Highway Research Board Special Report 29, NAS-NRC Publication 544: 20-47.

Varnes, D J 1978: 'Slope movement and types and processes.' In: Landslides: Analysis and Control. Edited by R L Schuster and R J Krizek, Transportation Research Board Special Report 176, National Academy of Sciences, Washington DC: 11-33.

Varnes, D J 1982: 'Methods of making landslide hazard maps.' In: Landslides and Mudflows: Reports of Alma-Ata International Seminar, October 1981. UNESCO/UNEP. Centre of International Projects, GKNT, Moscow: 388-406.

Waldron, L J; Dakessian, S 1981: Soil reinforcement by roots: calculation of increased soil shear resistance from root properties. Soil Science 132(6): 427-35.

Waltham, T 1978: Catastrophe: the Violent Earth. Macmillan Publishers.

Ward, T J; Li, R M; Simons, D B 1981: Use of a mathematical model for estimating potential landslide sites in steep forested drainage basins. Proceedings of Symposium on Erosion and Sediment Transport in Pacific Rim Steeplands, Christchurch, New Zealand 1981. Edited by T R H Davies and A J Pearce. International Association of Hydrological Sciences. Publication 132: 21-41.

Ward, W H 1945: The stability of natural slopes. Geographical Journal 105: 170-97.

Wegman, E 1957: Tectonique vivante, denudation et phenomenes connexes. Rev. Geog. Physique et Geol. Dynamique Pt. 2 1: 3-15.

Wellman, H W 1969: Tilted marine beach ridges at Cape Turakirae, New Zealand. Tuatara 17: 82-93.

Wellman, H W 1975: The obduction-subduction part of the Australian Pacific plate boundary in New Zealand. International Union of Geology and Geophysics, Commission on Recent Crustal Movements, Circular 10.

Wellman, H W; Wilson, A T 1965: Salt weathering, a neglected geological agent in coastal and arid environments. Nature 205: 1097-8.

Wells, W G 1981: Some effect of brushfires on erosion processes in coastal southern California. Proceedings of Symposium on Erosion and Sediment Transport in Pacific Rim Steeplands, Christchurch, New Zealand 1981. Edited by T R H Davies and A J Pearce. International Association of Hydrological Sciences. Publication 132: 305-42.

Whalley, W B; Douglas, G R; McGreevy, J P 1982: Crack propagation and associated weathering in igneous rocks. Zeitschrift fur Geomorphologie 26(1): 33-54.

White, G; Haas J 1975: Assessment of Research on Natural Hazards. MIT Press, Cambridge, Mass.

White, S E 1976: Is frost action really only hydration shattering? Arctic and Alpine Research 8: 1-6.

Wieczorek, G F; Sarmiento, J 1983: Significance of storm intensity-duration for triggering debris flows near La Honda, California. Geological Society of America Abstracts with Programs 15(5): 289.

Williams, R B G; Robinson, D A 1981: Weathering of sandstone by the combined action of frost and salt. Earth Surface Processes and Landforms 6: 1-9.

Winkler, E M 1977: Insolation warmed over: comment and reply. Geology 5: 188-90.

Winkler, E M; Wilhelm, E J 1970: Salt burst by hydration pressures in architectural stone in urban atmosphere. Proceedings of the Soil Science Society of America 25: 246-9.

Wright, A C S; Miller, R B 1952: Soils of southwest Fiordland. Soil Bureau Bulletin 7.

Wright, C; Mella, A 1963: Modifications to the soil pattern of South-Central Chile resulting from seismic and associated phenomena during the period May to August 1960. Bulletin Seismological Society of America 53: 1367-402.

Wu, T H; McKinnell, W P; Swanston, D N 1979: Strength of tree roots and landslides on Prince of Wales Island, Alaska. Canadian Journal of Geotechnical Research 16(1): 19-33.

Yatsu, E 1966: Rock Control in Geomorphology. Sozosha, Japan.

Yee, C S; Harr, R D 1977: Influence of soil aggregation on slope stability in the Oregon Coast Ranges. Environmental Geology 1: 367-77.

Young, A 1961: Characteristic and limiting slope angles. Zeitschrift fur Geomorphologie 5: 126-31.

Young, A 1972: Slopes. Longman.

Young, A 1974: The rate of slope retreat. Institute of British Geographers Special Publication 7: 65-78.

Young, A R M 1978: The influence of debris mantles and local climatic variations on slope stability near Wollongong, Australia. Catena 5: 95-107.

Zapata Luyo, M 1977: 'Origen y Evolucion de las Lagunas.' Instituto de Geologia y Mineria, Oficina Regional, Huaraz, Peru (Unpublished report in Spanish).

Zaruba, Q; Mencl, V 1969: Landslides and their Control. Elsevier/Academia, Prague.

Ziemer, R R 1981: Roots and the stability of forest slopes. Proceedings of Symposium on Erosion and Sediment Transport in Pacific Rim Steeplands, Christchurch, New Zealand 1981. Edited by T R H Davies and A J Pearce. International Association of Hydrological Sciences. Publication 132: 343-57.

Ziemer, R R; Swanston, D N 1977: Root strength changes after logging in southeast Alaska. United States Department of Agriculture Forest Service Research Note PNW-306.

SUBJECT INDEX

39-43, 47, 69, 77, 79, 87, 159

Responsibility 194, 195, 204

Return periods 181, 209

River erosion (see also fluvial process) 115-118

River load (see sediment yield)

Rock avalanches 52

Rock falls 77, 78

Rock mass strength 99, 102, 103, 104, 105, 118

Rock slides 29, 52, 60, 77, 101, 130, 131, 132, 170

Rock type 92, 93

Rocks 92, 99-104, 106

Root wedging 66-69

Roots 48, 66, 68, 82, 158, 159, 160

Rotational slides (see slumps)

Salts 71, 72; wedging 71-72

Sand 62, 63, 79

Sand dunes 58

Scree 84, 133

Sea-level (see eustatic changes)

Sediment concentrations 201

Sediment delivery ratio 118

Sediment storage 144, 153

Sediment transport 153, 155

Sediment transport, chronic 124

Sediment transport, pulse 124

Sediment yield 117, 118, 120, 123, 124, 125

Seepage 12, 62, 63, 81, 88

Seismicity 77-79, 113

Sensitive clays 79, 98

Sensitivity 97, 98

Shaved contact 147

Shear failure 24

Shear plane 11, 44, 59, 60, 87, 89, 99, 105, 107, 130, 134; angle of 41, 42, 47, 52, 53, 57

Shear strength (see also resistance) 32, 44, 53, 54, 80, 83, 90, 92, 94,

166, 167, 168; drained 13; effective 13; peak 12, 104, 105, 176; residual 12, 13, 105, 106, 128, 132; undrained 13

Shear stress 11, 12, 13, 32, 38, 39-43, 53, 60, 66, 77, 79

Shear surface (see shear plane)

Shearing resistance, angle of (see internal friction)

Shrinkage 63, 74

Shrinkage limit 74, 97

Silts 94

Slab failures 52

Slaking (see dispersibility)

Slide surface (see shear plane)

Slides 20

Slope adjustment 130, 131

Slope angle 13, 50, 53, 57-61, 130, 133, 149; critical (see limiting angle); stable (see limiting angle)

Slope aspect 62, 130, 162, 163, 164, 165, 166, 167, 168

Slope, convergent 140, 141

Slope development 35, 37

Slope, equilibrium 126

Slope erosion 117

Slope failure 7

Slope, form 90, 91

Slope height 11, 52-57, 112, 130; critical 52, 53, 55, 91, 107, 128

Slope instability 6; causes 32-110; concept 32-37; principles 37

Slope movement 7; flowage 13, 14, 20, 24, 50, 107; type 8, 107-110

Slope movements 7, 23, 24; classification (see classification); deep-seated (see landslides); depth 11; history 33, 35; material 8; morphometric characteristics 8, 3-31;

Slope-ripening 130, 131, 176